Optical Trapping and Manipulation

Optical Trapping and Manipulation

From Fundamentals to Applications

Editors

Philip Jones
Daniel R. Burnham

MDPI • Basel • Beijing • Wuhan • Barcelona • Belgrade • Manchester • Tokyo • Cluj • Tianjin

Editors
Philip Jones
Department of Physics and
Astronomy, University College
London
UK

Daniel R. Burnham
The Francis Crick Institute
UK

Editorial Office
MDPI
St. Alban-Anlage 66
4052 Basel, Switzerland

This is a reprint of articles from the Special Issue published online in the open access journal *Micromachines* (ISSN 2072-666X) (available at: https://www.mdpi.com/journal/micromachines/special_issues/optical_trapping_manipulation).

For citation purposes, cite each article independently as indicated on the article page online and as indicated below:

LastName, A.A.; LastName, B.B.; LastName, C.C. Article Title. *Journal Name* **Year**, *Article Number, Page Range*.

ISBN 978-3-03943-537-1 (Hbk)
ISBN 978-3-03943-538-8 (PDF)

Contents

About the Editors

Phil Jones studied at Cambridge University, Imperial College London and Oxford University, where he obtained his doctorate working on laser cooling and optical lattices. He is currently Professor of Physics at University College London (UCL), where he leads the Optical Tweezers Group. His research interests include optical trapping and optical binding and their application to biological, nanoscopic and soft materials. In 2015, he co-authored the book Optical Tweezers: Principles & Applications with Onofrio Maragò and Giovanni Volpe.

Daniel Burnham obtained his PhD from the University of St Andrews, having worked on holographic beam shaping and optically trapped aerosols. He held a Lindemann Trust Fellowship at the University of Washington, Seattle, followed by postdoctoral positions at the Delft University of Technology and The Francis Crick Institute, London, where he has also held a Marie Skłodowska-Curie Fellowship to work on single molecule imaging.

 *micromachines*

Article

Influence of Pulsed He–Ne Laser Irradiation on the Red Blood Cell Interaction Studied by Optical Tweezers

Ruixue Zhu [1], Tatiana Avsievich [1], Alexander Bykov [1], Alexey Popov [1] and Igor Meglinski [1,2,3,4,5,*]

[1]   Optoelectronics and Measurement Techniques Laboratory, University of Oulu, 90570 Oulu, Finland; ruixue.zhu@oulu.fi (R.Z.); tatiana.avsievich@oulu.fi (T.A.); alexander.bykov@oulu.fi (A.B.); alexey.popov@oulu.fi (A.P.)
[2]   Interdisciplinary Laboratory of Biophotonics, National Research Tomsk State University, 634050 Tomsk, Russia
[3]   Institute of Engineering Physics for Biomedicine (PhysBio), National Research Nuclear University (MEPhI), 115409 Moscow, Russia
[4]   Aston Institute of Materials Research, School of Engineering and Applied Science, Aston University, Birmingham B4 7ET, UK
[5]   School of Life and Health Sciences, Aston University, Birmingham B4 7ET, UK
*   Correspondence: i.meglinski@aston.ac.uk

Received: 4 November 2019; Accepted: 3 December 2019; Published: 5 December 2019

Abstract: Optical Tweezers (OT), as a revolutionary innovation in laser physics, has been extremely useful in studying cell interaction dynamics at a single-cell level. The reversible aggregation process of red blood cells (RBCs) has an important influence on blood rheological properties, but the underlying mechanism has not been fully understood. The regulating effects of low-level laser irradiation on blood rheological properties have been reported. However, the influence of pulsed laser irradiation, and the origin of laser irradiation effects on the interaction between RBCs remain unclear. In this study, RBC interaction was assessed in detail with OT. The effects of both continuous and pulsed low-level He–Ne laser irradiation on RBC aggregation was investigated within a short irradiation period (up to 300 s). The results indicate stronger intercellular interaction between RBCs in the enforced disaggregation process, and both the cell contact time and the initial contact area between two RBCs showed an impact on the measured disaggregation force. Meanwhile, the RBC aggregation force that was independent to measurement conditions decreased after a short time of pulsed He–Ne laser irradiation. These results provide new insights into the understanding of the RBC interaction mechanism and laser irradiation effects on blood properties.

Keywords: optical tweezers; red blood cells (RBCs); RBCs interactions; Helium–Neon laser; laser irradiation

1. Introduction

Human red blood cells (RBCs), or erythrocytes, can form face-to-face two-dimensional rouleaux, which can branch into three-dimensional structures by side-to-side or side-to-end contact of cells in aqueous suspension of large proteins or polymers. The spontaneous process of RBCs clumping is known as aggregation, whereas the enforced rouleaux dispersion by high-shear conditions is referred to as disaggregation. RBC (dis)aggregation highly depends on suspension properties including the concentration of fibrous biomolecules, the molecular weight of neutral polymers, and the antagonistic or synergetic effects between different proteins [1,2]. The (dis)aggregation behavior also relies on RBC intrinsic properties, such as cell age, shape, other donor-specific factors [3], membrane electrical properties (zeta potential) [4], RBC deformability [5], and properties of the glycocalyx on the outer surface of the cell membrane [6]. Measurements and analysis of RBCs mutual interaction have long been a hot topic of real-life significance for several reasons. First, RBC interaction is one of the most important hemorheological determinants that affects blood flow rheology, functional capillary density, organ perfusion, and distribution of other cells (e.g., leucocytes and platelets margination) Ref [7]. Secondly, the speed of blood aggregation is critical for hemostasis to control hemorrhage and maintain the circulating blood within the intravascular space when external or internal bleeding occurs [8]. Moreover, the degree of RBC aggregation is closely related to the pathological status of certain diseases (e.g., inflammation, cardiovascular embolism, and rheumatoid arthritis), and may serve as an early indicator for diagnosis [9]. Thus, the better understanding of RBC interaction mechanism and investigation of potential methods to regulate this process are important for the future blood microcirculation monitoring and therapy.

Typical RBC aggregation analyses are based on parameters indicating the extent and rate of rouleaux formation calculated from the time course required for RBC suspension to approach a final stable sedimentation state [10], individual cell-cell interaction dynamics cannot be revealed in detail. The development of single-cell level methods such as micropipette aspiration technique (MAT), atomic force microscopy (AFM), and scanning electron microscopy (SEM), has promoted the cell biology study and helped unravel the different structures and function of living cells on a microscopic and molecular level [11,12]. Among these methods is the Optical Tweezers (OT), a significant achievement of laser physics with the ability to non-invasively trap, manipulate and displace a living cell or part of it with highly accurate positioning that has deepened the investigation of intercellular interactions [13,14]. Since the first application of OT to RBC interaction study in 1997 [15], numerous significant results, such as quantified RBC parameters (e.g., membrane viscosity and zeta potential) [16] and the effects of heparin and tranexamic acid on the efficiency of RBC aggregation [17], have been obtained. Furthermore, based on the RBC aggregation behavior measured by OT, a sensitive monitoring method for systemic lupus erythematosus (SLE) and its response to drug therapies has been established [18]. The 2018 Nobel Prize in physics was awarded to the pioneer Dr Arthur Ashkin "for the optical tweezers and their application to biological systems". Unquestionably, the unique advantages of OT in micromanipulation as well as in precision measurement of ultra-low forces at piconewton (10^{-12} N) level have made it an essential tool in the field of biological cytology.

He–Ne laser sources are widely used in clinical diagnosis and treatments including accelerating wound healing of soft tissue, regulating metabolic activities, and aiding the treatment for acute cerebral infarction [19,20]. Low-level laser irradiation has great potentiality of regulating blood rheological properties (e.g., blood viscosity and electrophoretic mobility) and RBCs interaction in blood samples with abnormally high erythrocyte sedimentation rate (ESR) [21]. However, most studies were based on statistical analysis by traditional instruments such as Wintrobe tube for ESR, red cell deformability meter, and electrophoretic meter for zeta potential [22,23]. The origin of laser irradiation effects on blood properties, especially on RBC aggregation behavior is not clearly understood. The absorption of irradiated energy by hemoglobin is assumed to be the main mechanism as greater influence was observed by laser wavelength closer to the hemoglobin absorption band (400–600 nm) with a longer irradiation time (up to 130 min) [21,24].

In our study, OT was used to thoroughly assess RBC (dis)aggregation properties and investigate short-time laser irradiation effects on RBC interaction at a single-cell level. The results not only clarified the applicability of the two co-existent models to different RBC interaction processes, but also indicated the dependence of the disaggregation force on both the cell interaction time and initial contact area between two RBCs. Furthermore, by studying low-level laser irradiation effects on RBC aggregation within a short irradiation period (up to 300 s), an attempt was made to explore other possible laser-cell interaction mechanism rather than thermal absorption, which is predominant in long-time irradiation. The results showed that low-level He–Ne laser with appropriate pulse frequency has the potentiality to reduce the RBC aggregation force in a short irradiation time (120 s).

2. Materials and Methods

The experimental samples were highly diluted RBCs suspensions in platelet-low autologous plasma (hematocrit < 1%) donated by a single healthy donor (female, age 26) with oral consent and under ethical permission (Finnish Red Cross, No. 11/2019) to avoid introducing donor-specific differences to the measurements. The fresh RBCs were obtained by centrifuging diluted suspensions of fingertip-prick blood drops in phosphate buffer saline (PBS) at 6500 RPM (4732 g) for 10 min every time before experiments. The platelet-low plasma was obtained by double washing of whole blood samples obtained by venipuncture at a Nordlab clinic (Oulu, Finland) under the same condition, and was stored in a fridge for up to one month. The sample chamber with capacity of 20–30 μL liquid samples was composed of a microscope slide and a cover glass that were bonded by double-sided tape (100 μm thick), and was sealed by Vaseline after sample injection. All measurements were performed within 3 h after sample preparation at room temperature (23 °C).

Double-beam polarization-based optical tweezers (OT) was used as the main experimental tool to investigate cell interaction dynamics. The basic requirement is the strong focusing of the laser beam. The trapping phenomenon can be explained by momentum transfers involved in the interaction between the strong focused laser beam and the micro-sized particle (RBC) adjacent to the beam focus [25]. The construction of the in-house-made two-channel OT system combined with a chopper-modulated pulsed laser irradiation module is shown in Figure 1. Laser beam from an infrared single-mode Nd:YAG laser source (1064 nm, 350 mW, ILML3IF-300 Leadlight Technology, Taiwan) was divided into two orthogonally polarized continuous-wave beams that were then simultaneously focused by a large numerical aperture water immersion objective (lLUMPlanFl 100×/1.00 W, Olympus, Tokyo, Japan) to form two independent trapping channels in the focal plane within the sample chamber. One trapping beam was located in the center of the observation plane, and the position of the second trap was adjustable by stirring mirrors 1 and 2. The trapping power and the power ratio between the two channels were controlled through rotating half-wave plates. The orthogonal polarization states were adopted to reduce possible interference between the two beams, and the trapped cells were observed to align along the polarization direction within an optical trap. A beam expander was used to widen the trapping beams to fill the back aperture of the objective to achieve the maximum trapping efficiency. The observation system was made up of a white LED-illuminating source, a dichroic mirror, an IR-filter, and a CMOS camera (lPixelink PL-B621M, Ottawa, Ontario, Canada). The laser irradiation module consists of a He–Ne laser source (633 nm, 4 mW, beam diameter 0.48 mm, 1507P-0, JDS Uniphase, USA) and an optical chopper system (MC2000B, Thorlabs, USA) was used to irradiate the trapped RBC rouleau in the sample chamber. As shown in Figure 1, the linear overlapping distance x between the two trapped RBCs can be measured from the captured top-view image, and the corresponding interaction area A, which is enclosed by two arcs s (marked in yellow in the side view image), can be calculated as the sum of two segments using the following equation:

$$A = R^2 \cdot (\theta - \sin\theta),\tag{1}$$

where R is the radius of the RBC and θ is the central angle of the arc:

$$\theta = 2 \cdot \arccos\left(\frac{R-h}{h}\right) = 2 \cdot \arccos\left(\frac{R-x/2}{x/2}\right),\tag{2}$$

where h is the height of the arc (the perpendicular from the midpoint of the arc's chord to the arc itself). If the interaction force F_{inter} is known under this overlapping area A, the corresponding interaction energy density U between the two RBCs can be estimated by the following equation:

$$U = \frac{F_{inter} \cdot 2h}{A}.\tag{3}$$

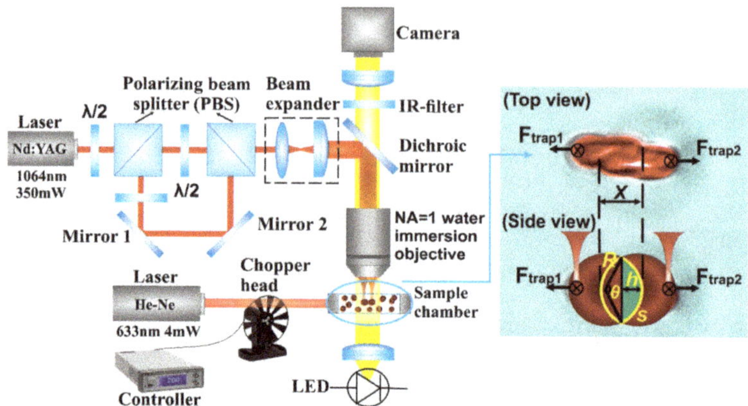

Figure 1. Schematic layout of the double-channel optical tweezers system combined with a chopper-modulated pulsed laser irradiation module for studying RBC interaction and laser irradiation effects. Two orthogonally polarized beams separated from one infrared laser source were focused simultaneously to form two independent trapping channels. He–Ne laser modulated by an optical chopper system was used to irradiate the trapped RBC rouleau in the sample chamber. Top-view image was captured by the CMOS camera in transmission mode and the interaction area between the two RBCs was calculated according to the side view geometries.

To quantify the optical trapping force applied to RBCs, force calibration was performed based on the force equilibrium between the optical trapping force and the viscous friction force exerted on a trapped RBC by a counterflow with a known velocity. In our case, blood plasma was the trapping environment, and the flow was generated by the lateral movement of the sample chamber attached to a XYZ-stage. The optical trapping force is linearly related to the trapping power, whereas the viscous friction force F_v is linearly related to the flow velocity v, as described by Stokes' law [26]:

$$F_v = 6\pi\eta RvK,\tag{4}$$

where η (1.30 mPas) is viscosity coefficient of the fluid (blood plasma) [27]; R (2.7 μm) is an effective radius of the RBC; K (1.36) is a correction factor for ellipsoid. The flow velocity was calculated from the video captured during the measurement. With a given trapping power, the velocity of the flow was slowly increased until the trapping force was matched by the viscous friction force and the RBC escaped from the trap.

3. Results

3.1. Trapping Force Calibration

The linear relationship between the optical trapping force and trapping power for each of the two trapping channels is shown in Figure 2. It can be clearly seen that the trapping force of the movable beam (trap 2) is slightly weaker than that of the central beam (trap 1). This was also manifested by the experimental phenomenon that RBCs escaped first from the movable trap. Therefore, the linear fitting equation of the movable channel $F_{trap2} = 0.11 \times P - 0.67$, where P is the optical trapping power, was used to calculate the optical trapping force. By measuring the laser power after the objective, the attenuation coefficient of the objective was calculated to be about 47%. The trapping efficiency Q was about 0.05, according to the following equation [28]:

$$Q = \frac{c \cdot F_{trap}}{P_e \cdot n_m},\tag{5}$$

where c is light propagation speed in vacuum, P_e is the effective trapping power, and n_m is the refractive index of the trapping medium.

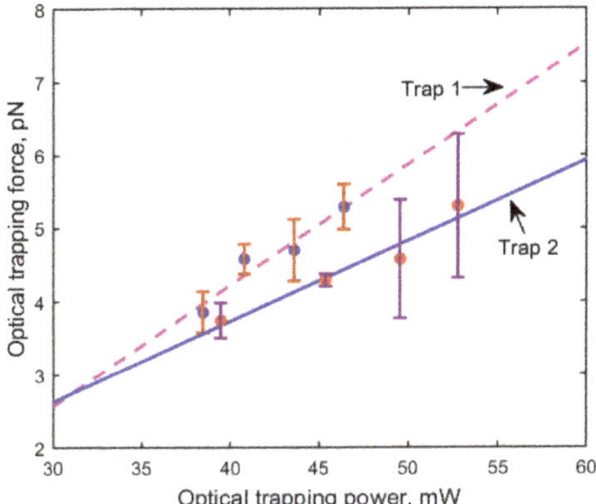

Figure 2. Trapping force calibration for the central beam (trap 1: pink dashed line) and for the movable beam (trap 2: black solid line). At least six measurements were performed with each trapping power. The trapping force in trap 2 is slightly weaker than in trap 1.

3.2. Interaction Energy Density of RBCs during Aggregation and Disaggregation Process

With the two-channel OT system, the intercellular forces between two RBCs in rouleau formation (aggregation) and destruction (disaggregation) were measured separately, with the interaction energy density calculated for each process. For aggregation measurement as shown in Figure 3a, two individual RBCs were separately captured and lifted to a height of 40 μm from the bottom of the sample chamber by two trapping beams simultaneously, and then a linear contact between the two cells was formed under the control of OT. After 40 s interaction, the aggregation force (F_A) was measured as the minimum optical force applied to stop the two cells from clumping together by slowly decreasing the trapping power while keeping the contact area constant.

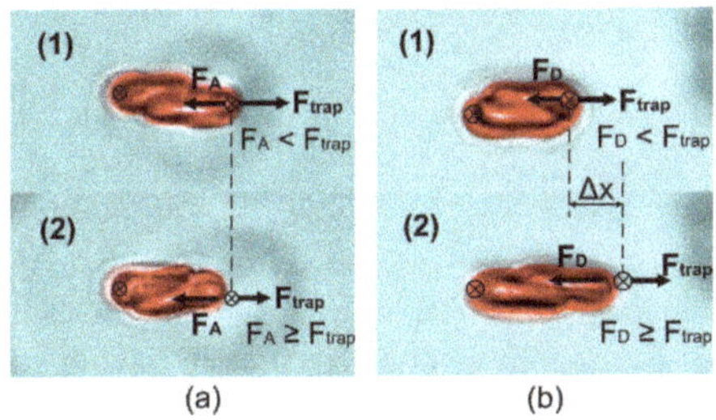

Figure 3. (**a**) Illustration of RBC aggregation force measurement. The position of the second trap was kept still while the trapping power was slowly decreased until the RBC escaped the trap; (**b**) Illustration of RBC disaggregation force measurement. The position of the second trap was slowly moved towards the direction to separate the two RBCs with a certain trapping power until the RBC escaped the trap. The small circles with crosses indicate the positions of the traps. F_{trap} indicates the trapping force, F_A and F_D indicate aggregation and disaggregation force respectively.

Among the two co-existent models, the "depletion layer model", which interprets the RBC clumping as the osmotic pressure caused by a depletion layer of low macromolecule concentration near the interaction surface between two RBCs, has been widely used to describe the aggregation mechanism [29,30]. Aggregation forces of 80 pairs of RBCs were measured with different initial linear overlapping distance (interaction area) in three independent groups to test the repeatability of the results. The corresponding interaction energy densities were calculated and plotted in Figure 4a. The measured aggregation force (F_A) is very small (in a range of 0.2–4.0 pN), so as the calculated interaction energy density (in a range of 0–0.9 aJ/μm^2). The measurement of RBC aggregation was sensitive to a variety of factors including the interference of the measurement environment and operation errors, thus the fluctuations in the measured values are evident. Despite the fluctuations, compared to the "bridging model", the relationship between the interaction energy density and the relative interaction area measured during RBC aggregation is more consistent with the linear description of the "depletion layer model". Therefore, a linear fitting was applied to the measured data as shown in Figure 4a. However, as suggested by a recent study [31], the "cross-bridging" mechanism is partly involved in RBC aggregation, thus the relationship may not be accurately described by one single model.

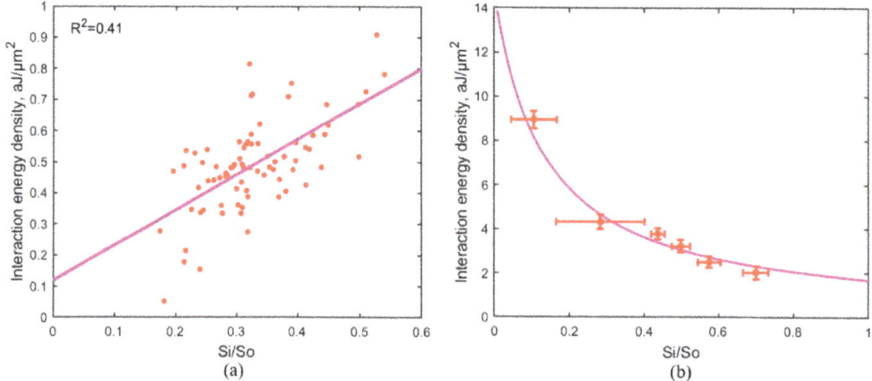

Figure 4. (**a**) RBC interaction energy density measured in aggregation process. 80 pairs of RBCs were measured with different initial overlapping in three independent groups. Experimental data and a linear fitting of the data are shown by red dots and a solid line respectively; (**b**) RBC interaction energy density measured in disaggregation process. 56 pairs of RBCs were measured in three independent groups with six different trapping strength. The reciprocal relationship between the interaction energy density and the relative interaction area is consistent with the "bridging model" prediction.

For disaggregation measurement as shown in Figure 3b, after 40 s cell contact at an 80% overlapping area, a constant optical trapping force was applied to slowly drag one RBC away from the other cell until at least one RBC escaped from the trap. The applied optical force was considered to be the disaggregation force (F_D) to decrease the contact area between the two cells. The interaction energy density reserved in the final interaction area between the two cells at the escaping moment was calculated using Equation (3). The "bridging model" assuming the two RBCs in contact are connected by "bridges" formed by physisorption of macromolecules is widely applied to interpret the intercellular interaction in disaggregation process [30,32]. 56 pairs of RBCs were measured in three independent groups with six different effective trapping power (trapping force) varied from 16.8 mW (3.3 pN) to 36.7 mW (8.0 pN). The obtained interaction energy density varied from 2.0 aJ/μm^2 to 9.0 aJ/μm^2. A reciprocal relationship between the interaction energy density and the relative interaction area was observed as shown in Figure 4b. According to the "bridging model", this relationship follows the expression [32]:

$$D = 2k_B T m_0 b \frac{1}{1 + b(S_i/S_0)},\tag{6}$$

where D is the interaction energy density and (S_i/S_0) is the relative interaction area; the product of Boltzmann constant and the absolute temperature $k_B T$ (4.11×10^{-21} J) is used as an energy value scaling factor. m_0 and b denoting the cross-bridge density and binding affinity are characteristic parameters of this model, and were calculated to be $1/4260\ \mathrm{nm}^{-2}$ (m_0) and 7.6 (b) in our measurement.

The result indicates that more interaction energy was accumulated within a smaller contact area between the two RBCs during the disaggregation process, which could be explained by the "bridging model" as more "bridges" connecting the two cells need to be broken to separate the two RBCs to a greater extent (to achieve a smaller final contact area). On the other hand, the reciprocal relationship can be explained as the "bridges" were not broken proportionally with the decreasing of contact area, but were sliding towards the conjugated area while remaining attached to both cells during the disaggregation process [2].

3.3. Effects of Short-Time Pulsed He–Ne Laser Irradiation on RBC Aggregation

Short irradiation time (up to 300 s) was adopted in our study to minimize thermal effects. As previously reported by us [33], the RBC aggregation dynamics is independent of cell contact time, whereas the measured disaggregation force is strongly influenced by the cell interaction time. With the aforementioned method, different final adhering areas between two RBCs were recorded during the enforced separation by the same trapping force of 9.2 pN after different cell adhering time (0–300 s) at the 80% initial overlapping. The result as shown in Figure 5 indicates that the longer the two RBCs adhere to each other, the harder to separate them from each other. It reveals the time-dependence of the "bridges" formation between two RBCs in the "bridging model".

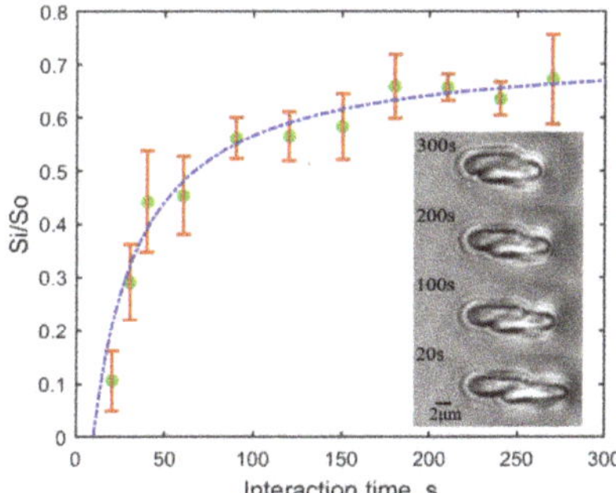

Figure 5. Time-dependence of the final relative adhering area (S_i/S_0) recorded after an enforced separation process by a trapping force of 9.2 pN from an 80% initial contact after different adhering time. 55 pairs of RBCs in total were measured (at least 3 pairs of RBCs for each condition). The green dots show the experimental data with standard deviations and the dashed curve denotes a possible trend of the influence. Microscopic images were captured at the moment of cell escaping, and the time denotes the contact time between the two RBCs before undergoing the disaggregation measurement.

Considering the influence of cell interaction time, the investigation of low-level laser irradiation effects was performed with the RBC aggregation process. An initial relative contact of 40% was formed between two RBCs and the continuous laser irradiation was started at the same time. The irradiation from the He–Ne laser source (power density: 2.21 W/cm^2) was directly applied to the trapped RBC rouleau. The RBC aggregation force was measured after different irradiation time. Figure 6 shows the aggregation forces measured in both the irradiated group (pink triangles) and the control group (without laser irradiation) (green dots). The RBC aggregation force was 2.0 ± 0.7 pN in the irradiated group, and 2.0 ± 0.5 pN in the control group. No significant influence of continuous laser irradiation (0–300 s) on RBC aggregation was observed in plasma.

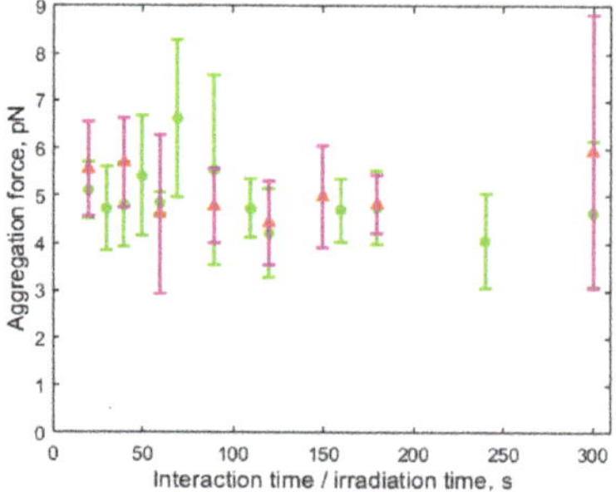

Figure 6. Aggregation forces measured with 32 pairs of RBCs with an initial relative contact area (S_i/S_0) of around 40% over different time of continuous He–Ne laser irradiation (0–300 s) (pink triangles), and the forces measured with 39 pairs of RBCs in the control group (green dots). No significant influence of continuous laser irradiation (within 300 s) on RBC aggregation was observed.

By applying frequency modulation to the continuous-wave He–Ne laser beam with an optical chopper, the effect of pulsed laser irradiation on RBC aggregation was studied. The RBC aggregation force was measured using the aforementioned method under pulsed laser irradiation with frequencies varied from 25 Hz to 10,000 Hz at step size of 25 Hz. The irradiation was applied directly to the trapped RBC rouleau for 60 s, 120 s, and 180 s separately, and the results were compared with the control group (without irradiation). Among all the irradiation conditions, a decrease in RBC aggregation force was observed with 225 Hz pulsed irradiation for 120 s as shown in Figure 7, in which the negative value of aggregation force indicates the two RBCs did not aggregate. Experiments were conducted with 50 pairs of RBCs for control in two groups, and 50 pairs of RBCs for pulsed laser irradiation in three groups. The difference between any of the irradiation groups and any of the control groups is statistically significant ($p < 0.05$) examined by Mann-Whitney U test.

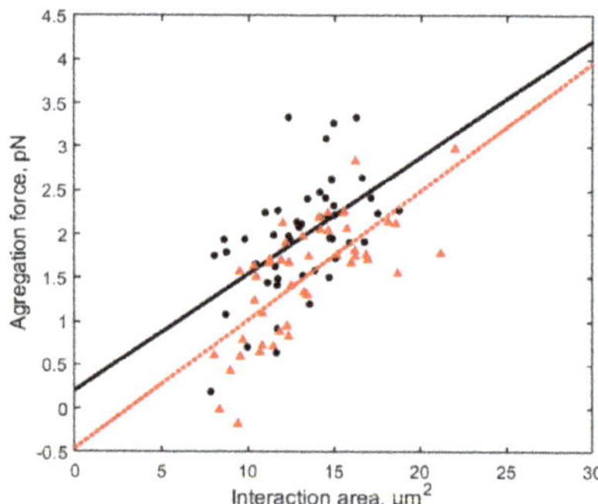

Figure 7. RBC aggregation forces measured with 120 s pulsed He–Ne laser irradiation with pulse frequency of 225 Hz (red triangles, 50 pairs of RBCs) and without irradiation (control group) (black dots, 50 pairs of RBCs). The negative value of aggregation force indicates the two RBCs did not aggregate.

4. Discussion

The RBC dis/aggregation force measured in the present study are consistent with those obtained earlier by Prof A.V. Priezzhev's group [2,34], whereas the disaggregation energy density is smaller than previously reported by our group [12]. Nevertheless, the obtained characteristic parameters of the "bridging model" ($b = 7.6$ and $m_0 = 1/4260 \ \text{nm}^{-2}$) are consistent with our group's previous study. This can be explained by the dependence of the measured disaggregation energy density on the initial contact area between two RBCs, as shown in Figure 8. It can be seen that larger initial overlapping area leads to higher measurable interaction energy density between two RBCs in the disaggregation measurement. This phenomenon can be interpreted by the "bridging model" as more "bridges" were formed within a larger initial overlapping area, and in order to separate the two cells, more "bridges" needed to be broken compared with a smaller initial overlapping. Thus, stronger disaggregation force and higher interaction energy were measured when the initial overlapping area is larger. A similar observation was also reported by Prof A.V. Priezzhev's group [35] by measuring the disaggregation forces corresponding to a 20% relative contact area under a 20% or 80% initial overlapping.

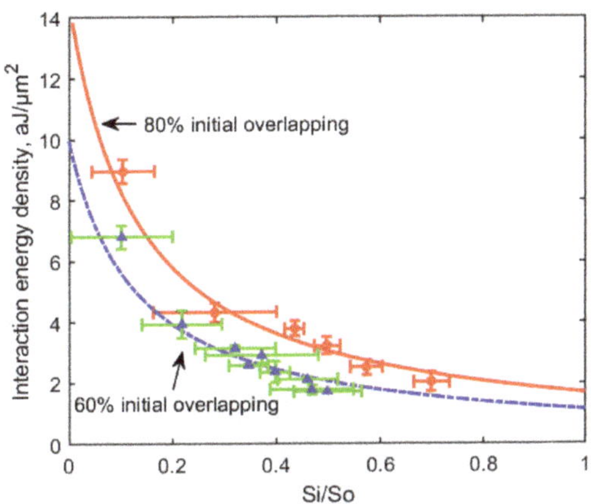

Figure 8. Relationship curves between the RBC disaggregation energy density and relative interaction area measured with 80% (red dots: experimental data, solid line: theory) and 60% (blue triangles: experimental data, dashed line: theory) initial overlapping area. Larger initial overlapping area leads to higher measurable disaggregation interaction energy density between two RBCs.

Among existing studies on low-level He–Ne laser irradiation effects on blood rheological properties, most results were observed after long-time irradiation (20–130 min) [21–24], and the thermal effect originating from irradiation energy absorption by hemoglobin was assumed to be the main mechanism. The present study shows that low-level pulsed He–Ne irradiation with pulse frequency of 225 Hz has the potentiality to decreased the RBC aggregation force after a short time of irradiation (120 s). This indicates that other mechanism, such as the modulation of the vibrational state of RBC membrane or membrane-attached proteins might be involved during short-time laser irradiation. Such hypothesis is supported by another study indicating the improvement in RBC deformability after laser irradiation is caused by the detaching of membrane-bound hemoglobin from red cell membrane due to increased vibration energy [21]. However, the systematic evaluation of pulsed laser irradiation effects on RBC interaction dynamics, as well as the detailed mechanism involved in interaction between laser irradiation and RBCs require further in-depth investigation.

As with the majority of studies [12,30,31], the presented results were obtained based on the analysis of the behavior of RBCs taken from one donor, and all measurements were conducted in autologous plasma, thus the results may be specific for the blood of the recruited donor.

5. Conclusions

Red blood cell (dis)aggregation properties were investigated in detail by optical tweezers. A linear and a reciprocal dependence of the interaction energy density on the relative interaction area between two RBCs in autologous plasma were obtained in aggregation and disaggregation process, respectively. The interaction energy measured in RBC disaggregation is influenced by the measurement conditions (e.g., initial cell overlapping area and contact time) and has different orders of magnitude than that measured in RBC aggregation. In addition, up to 300 s continuous He–Ne laser irradiation (power density: 2.21 W/cm^2) showed no influence on RBC aggregation, whereas the RBC aggregation force decreased after 120 s irradiation by the same laser with pulse modulation of 225 Hz. The observation indicates the potentiality to modulate RBC aggregation properties by short-time laser irradiation. Our study provides new insights into the investigation of the dynamic interaction between RBCs, and the evaluation of the regulating effects of low-level laser irradiation on RBC interaction properties.

Author Contributions: Conceptualization, I.M. and R.Z.; Methodology, R.Z. and T.A.; Investigation, R.Z.; Funding acquisition, I.M.; Writing—original draft preparation, R.Z.; Writing—review and editing, I.M., T.A., R.Z.; Supervision, I.M., A.B. and A.P.

Funding: This research was partially supported by China Scholarship Council (CSC No. 201706410089, R.Z.), Tauno Tönning Foundation (grant No. 20190104, R.Z.), EDUFI Fellowship (TM-17-10370, TM-18-10820, T.A.), Suomen Kulttuurirahasto (grant No. 00190188, T.A.), Oulun yliopiston apuraharahasto (No. 20180107, T.A.). I.M. also acknowledges partial support from the Academy of Finland (project 326204, 325097), MEPhI Academic Excellence Project (Contract No. 02.a03.21.0005), and the National Research Tomsk State University Academic D.I. Mendeleev Fund Program.

Conflicts of Interest: The authors declare no conflict of interest.

References

1. Pribush, A.; Zilberman-Kravits, D.; Meyerstein, N. The mechanism of the dextran-induced red blood cell aggregation. *Eur. Biophys. J.* **2007**, *36*, 85–94. [CrossRef] [PubMed]
2. Lee, K.; Kinnunen, M.; Khokhlova, M.D.; Lyubin, E.V.; Priezzhev, A.V.; Meglinski, I.; Fedyanin, A.A. Optical tweezers study of red blood cell aggregation and disaggregation in plasma and protein solutions. *J. Biomed. Opt.* **2016**, *21*, 035001. [CrossRef] [PubMed]
3. Rampling, M.; Meiselman, H.; Neu, B.; Baskurt, O. Influence of cell-specific factors on red blood cell aggregation. *Biorheology* **2004**, *41*, 91–112. [PubMed]
4. Jan, K.M.; Chien, S. Role of surface electric charge in red blood cell interactions. *J. Gen. Physiol.* **1973**, *61*, 638–654. [CrossRef] [PubMed]
5. Dao, M.; Lim, C.T.; Suresh, S. Mechanics of the human red blood cell deformed by optical tweezers. *J. Mech. Phys. Solids* **2003**, *51*, 2259–2280. [CrossRef]
6. Meiselman, H.J. Red blood cell aggregation: 45 years being curious. *Biorheology* **2009**, *46*, 1–19. [CrossRef]
7. Baskurt, O.K.; Meiselman, H.J. Hemodynamic effects of red blood cell aggregation. *Indian J. Exp. Biol.* **2007**, *45*, 25–31.
8. Litvinov, R.I.; Weisel, J.W. Role of red blood cells in haemostasis and thrombosis. *ISBT Sci. Ser.* **2017**, *12*, 176–183. [CrossRef]
9. Baskurt, O.K.; Meiselman, H.J. Erythrocyte aggregation: Basic aspects and clinical importance. *Clin. Hemorheol. Microcirc.* **2013**, *53*, 23–37. [CrossRef]
10. Baskurt, O.K.; Uyuklu, M.; Ulker, P.; Cengiz, M.; Nemeth, N.; Alexy, T.; Shin, S.; Hardeman, M.R.; Meiselman, H.J. Comparison of three instruments for measuring red blood cell aggregation. *Clin. Hemorheol. Microcirc.* **2009**, *43*, 283–298. [CrossRef]
11. Brehm-Stecher, B.F.; Johnson, E.A. Single-cell microbiology: Tools, technologies, and applications. *Microbiol. Mol. Biol. R.* **2004**, *68*, 538–559. [CrossRef] [PubMed]
12. Avsievich, T.; Popov, A.; Bykov, A.; Meglinski, I. Mutual interaction of red blood cells assessed by optical tweezers and scanning electron microscopy imaging. *Opt. Lett.* **2018**, *43*, 3921–3924. [CrossRef] [PubMed]
13. Ashkin, A. Acceleration and trapping of particles by radiation pressure. *Phys. Rev. Lett.* **1970**, *24*, 156. [CrossRef]
14. Ashkin, A.; Dziedzic, J.M. Optical trapping and manipulation of viruses and bacteria. *Science* **1987**, *235*, 1517–1520. [CrossRef]
15. Bronkhorst, P.; Grimbergen, J.; Brakenhoff, G.; Heethaar, R.; Sixma, J. The mechanism of red cell (dis) aggregation investigated by means of direct cell manipulation using multiple optical trapping. *Br. J. Haematol.* **1997**, *96*, 256–258. [CrossRef]
16. Fontes, A.; Fernandes, H.P.; de Thomaz, A.A.; Barbosa, L.C.; Barjas-Castro, M.L.; Cesar, C.L. Measuring electrical and mechanical properties of red blood cells with double optical tweezers. *J. Biomed. Opt.* **2008**, *13*, 014001. [CrossRef]

17. Yang, B.W.; Li, Z. Measuring micro-interactions between coagulating red blood cells using optical tweezers. *Biomed. Opt. Express* **2010**, *1*, 1217–1224. [CrossRef]

18. Khokhlova, M.; Lyubin, E.V.; Zhdanov, A.G.; Fedyanin, A.A.; Rykova, S.Y.; Sokolova, I.A. Normal and system lupus erythematosus red blood cell interactions studied by double trap optical tweezers: Direct measurements of aggregation forces. *J. Biomed. Opt.* **2012**, *17*, 025001. [CrossRef]

19. Yova, D.; Haritou, M.; Koutsouris, D. Antagonistic effects of epinephrine and helium-neon (He-Ne) laser irradiation on red blood cells deformability. *Clin. Hemorheol. Microcirc.* **1994**, *14*, 369–378. [CrossRef]

20. Zhu, J.; Liang, M.Y.; Cao, H.C.; Li, X.Y.; Li, S.M.; Li, S.H.; Li, W.Q.; Zhang, J.H.; Liu, L.; Lai, J.H. Effect of intravascular irradiation of He-Ne laser on cerebral infarction: Hemorrheology and apoptosis. In *Shanghai International Conference on Laser Medicine and Surgery*; Zhu, J., Ed.; SPIE: Bellingham, WA, USA, 2006; Volume 5967, p. 59671K.

21. Mi, X.; Chen, J.; Cen, Y.; Liang, Z.; Zhou, L. A comparative study of 632.8 and 532 nm laser irradiation on some rheological factors in human blood in vitro. *J. Photochem. Photobiol. B* **2004**, *74*, 7–12. [CrossRef]

22. Al Musawi, M.S.; Jaafar, M.; Al-Gailani, B.; Ahmed, N.M.; Suhaimi, F.M.; Bakhsh, M. Erythrocyte sedimentation rate of human blood exposed to low-level laser. *Laser Med. Sci.* **2016**, *31*, 1195–1201. [CrossRef] [PubMed]

23. Mi, X.Q.; Chen, J.Y.; Liang, Z.J.; Zhou, L.W. In vitro effects of helium-neon laser irradiation on human blood: blood viscosity and deformability of erythrocytes. *Photomed. Laser Surg.* **2004**, *22*, 477–482. [CrossRef] [PubMed]

24. Siposan, D.G.; Lukacs, A. Effect of low-level laser radiation on some rheological factors in human blood: An in vitro study. *J. Clin. Laser Med. Sur.* **2000**, *18*, 185–195. [CrossRef] [PubMed]

25. Nieminen, T.A.; Knöner, G.; Heckenberg, N.R.; Rubinsztein-Dunlop, H. Physics of optical tweezers. *Method. Cell Biol.* **2007**, *82*, 207–236.

26. Maklygin, A.Y.; Priezzhev, A.V.; Karmenian, A.; Nikitin, S.Y.; Obolenskii, I.; Lugovtsov, A.E.; Li, K. Measurement of interaction forces between red blood cells in aggregates by optical tweezers. *Quantum Electron.* **2012**, *42*, 500. [CrossRef]

27. Windberger, U.; Bartholovitsch, A.; Plasenzotti, R.; Korak, K.; Heinze, G. Whole blood viscosity, plasma viscosity and erythrocyte aggregation in nine mammalian species: Reference values and comparison of data. *Exp. Physiol.* **2003**, *88*, 431–440. [CrossRef]

28. Ashkin, A. Forces of a single-beam gradient laser trap on a dielectric sphere in the ray optics regime. *Biophys. J.* **1992**, *61*, 569–582. [CrossRef]

29. Neu, B.; Meiselman, H.J. Depletion-mediated red blood cell aggregation in polymer solutions. *Biophys. J.* **2002**, *83*, 2482–2490. [CrossRef]

30. Avsievich, T.; Popov, A.; Bykov, A.; Meglinski, I. Mutual interaction of red blood cells influenced by nanoparticles. *Sci. Rep.* **2019**, *9*, 5147. [CrossRef]

31. Lee, K.; Wagner, C.; Priezzhev, A.V. Assessment of the "cross-bridge"—Induced interaction of red blood cells by optical trapping combined with microfluidics. *J. Biomed. Opt.* **2017**, *22*, 091516. [CrossRef]

32. Tozeren, A.; Sung, K.; Chien, S. Theoretical and experimental studies on cross-bridge migration during cell disaggregation. *Biophys. J.* **1989**, *55*, 479–487. [CrossRef]

33. Zhu, R.; Avsievich, T.; Popov, A.; Meglinski, I. Influence of interaction time on the red blood cell (dis) aggregation dynamics in vitro studied by optical tweezers. In *Novel Biophotonics Techniques and Applications V*; SPIE: Bellingham, WA, USA, 2019; Volume 11075, p. 110750D.

34. Lee, K.; Kinnunen, M.; Danilina, A.V.; Ustinov, V.; Shin, S.; Meglinski, I.; Priezzhev, A.V. Characterization at the individual cell level and in whole blood samples of shear stress preventing red blood cells aggregation. *J. Biomech.* **2016**, *49*, 1021–1026. [CrossRef] [PubMed]

35. Lee, K.; Danilina, A.V.; Kinnunen, M.; Priezzhev, A.V.; Meglinski, I. Probing the red blood cells aggregating force with optical tweezers. *IEEE J. Sel. Top. Quantum* **2015**, *22*, 365–370. [CrossRef]

 *micromachines*

Article

Impact of Nanocapsules on Red Blood Cells Interplay Jointly Assessed by Optical Tweezers and Microscopy

Tatiana Avsievich [1,*], Yana Tarakanchikova [1,2,3], Ruixue Zhu [1], Alexey Popov [1,*], Alexander Bykov [1], Ilya Skovorodkin [4], Seppo Vainio [4,5] and Igor Meglinski [1,6,7,8,9,*]

1   Optoelectronics and Measurement Techniques Research Unit, University of Oulu, 90014 Oulu, Finland; yana.tarakanchikova@oulu.fi (Y.T.); ruixue.zhu@oulu.fi (R.Z.); alexander.bykov@oulu.fi (A.B.)
2   Nanobiotechnology Laboratory, St. Petersburg Academic University, St. Petersburg 194021, Russia
3   RASA center in St. Petersburg, Peter the Great St. Petersburg Polytechnic University, St. Petersburg 195251, Russia
4   Biocenter Oulu and Faculty of Biochemistry and Molecular Medicine, Biocenter Oulu, Laboratory of Developmental Biology, University of Oulu, 90014 Oulu, Finland; ilya.skovorodkin@oulu.fi (I.S.); seppo.vainio@oulu.fi (S.V.)
5   InfoTech Oulu, Borealis Biobank of Northern Finland, Oulu University Hospital, University of Oulu, 90014 Oulu, Finland
6   Interdisciplinary Laboratory of Biophotonics, National Research Tomsk State University, Tomsk 634050, Russia
7   Institute of Engineering Physics for Biomedicine (PhysBio), National Research Nuclear University MEPhI, Moscow 115409, Russia
8   Aston Institute of Materials Research, School of Engineering and Applied Science, Aston University, Birmingham B4 7ET, UK
9   School of Life & Health Sciences, Aston University, Birmingham B4 7ET, UK
*   Correspondence: tatiana.avsievich@oulu.fi (T.A.); alexey.popov@oulu.fi (A.P.); i.meglinski@aston.ac.uk (I.M.)

Received: 16 November 2019; Accepted: 17 December 2019; Published: 23 December 2019

Abstract: In the framework of novel medical paradigm the red blood cells (RBCs) have a great potential to be used as drug delivery carriers. This approach requires an ultimate understanding of the peculiarities of mutual interaction of RBC influenced by nano-materials composed the drugs. Optical tweezers (OT) is widely used to explore mechanisms of cells' interaction with the ability to trap non-invasively, manipulate and displace living cells with a notably high accuracy. In the current study, the mutual interaction of RBC with polymeric nano-capsules (NCs) is investigated utilizing a two-channel OT system. The obtained results suggest that, in the presence of NCs, the RBC aggregation in plasma satisfies the 'cross-bridges' model. Complementarily, the allocation of NCs on the RBC membrane was observed by scanning electron microscopy (SEM), while for assessment of NCs-induced morphological changes the tests with the human mesenchymal stem cells (hMSC) was performed. The combined application of OT and advanced microscopy approaches brings new insights into the conception of direct observation of cells interaction influenced by NCs for the estimation of possible cytotoxic effects.

Keywords: polymeric nanocapsules; red blood cells; optical tweezers; SEM; optical microscopy; cytotoxicity; human mesenchymal stem cells

---

1. Introduction

Nowadays, there are numerous delivery systems such as: natural biological vehicles (such as bacteria and viruses [1,2], various cell types including red blood cells (RBCs) [3–5], immune cells [6,7], stem cells [8–11] and manufactured carriers (liposomes [12], micelle [13], polymeric capsules [14–17], polymeric complexes [18–21]). Polymeric nanocapsules (NCs) are a promising system of natural carriers for targeted drug delivery because of their low cytotoxicity, shell flexibility, and high in vivo stability [22–26]. A wide range of drugs [27–29] and nucleotides of ribonucleic acid (RNA)/deoxyribonucleic acid (DNA) [30–32] can be effectively encapsulated in the polymer capsules. Effective distribution of drugs in target tissues can be achieved by active delivery methods that are potentially much more beneficial than passive accumulation. This is why design of new drug carriers is one of the fastest developing directions of research in the nanomedicine area. This allows, in particular, creating a controllable photosensitizer delivery system for cell applications [33–35]. Increasing use of nanomaterials, including nanoparticles and NCs in household, medical and industrial products requires a careful assessment of possible consequences [36–38].

A promising area of targeted delivery of physiologically active substances implies the use of polymeric NCs for transport of medical drugs through the blood vessels. However, there is a potential risk of negative impact of NCs on blood cells [39]. RBCs, being the most abundant blood component, easily obtainable and available, remain one of the most common test systems for characterizing the action of biologically active compounds at the cellular and subcellular levels [40]. For administration of NCs into a human organism the bloodstream serves as a common pathway for delivering medicine (including nanodrugs), but it remains largely unknown and hardly predictable, whether nanoparticles affect blood components. Intravenous injection assumes circulation of NCs in a blood stream for a certain time. During this period, the injected drugs should be delivered to targeted tissues with minimized adverse influence or toxic effects on organs and systems used as a delivery pathway.

The understanding of NCs' behavior in close to natural conditions will help to predict the possible complications and undesirable side effects of NC-based drug delivery systems. Another promising direction is the development of hybrid delivery systems, when a cell is coupled with an active component [41,42], e.g., NCs, thus increasing the effectiveness of the delivery and ensuring prolonged circulation of the drug in the organism. Therefore, revealing the influence of NCs on cells a single-cell level, can bring new insights, thus helping to improve biocompatibility of NCs.

RBC aggregation is a reversible process of adhering RBCs to each other in a face-to-face morphology, which depends on the blood plasma composition (macromolecules, including plasma proteins) and RBC cellular properties [43]. Aggregation of RBCs affects blood viscosity and plays a crucial role in the blood microcirculation [44]. Its characteristics are sensitive to the majority of socially significant diseases, such as diabetes, inflammation, cardiovascular embolism, myocardial ischemia etc. [45], hence RBC aggregation state can be used as a marker of the organism physiological status. In the present research RBC aggregation is evaluated through the mutual RBCs interactions between the individual single cells.

The optical tweezers (OT) technique allows for manipulation of microscopic objects including cells with a strongly focused laser beam and for measuring interaction forces between them within 0.1–100 pN. The method was first introduced by A. Ashkin [46,47], who was awarded the Nobel Prize in Physics for this development in 2018. OT method has been applied for a variety of applications in the field of physics and biology. The method was efficiently used in RBCs studies for measuring aggregation forces between individual interacting cells and to demonstrate possibility of performing experiments in the natural state of RBCs without impairing their properties. A recent study demonstrates successful measurements of such forces in autologous plasma and in protein solutions [48].

The present study focuses on the evaluation of RBC response to NCs in terms of RBC aggregation in autologous plasma using OT, complemented with SEM of RBCs. Additionally, microscopic observations of changes in morphology of human mesenchymal stem cells (hMSCs) were performed. HMSCs represent a significantly different type of cells compared to RBCs, being less deformable, phagocytic (able to engulf other cells or particles) and having a nucleus. A recent study [42] demonstrated that hMSCs cells coupled with NCs can serve as an effective antitumor drug delivery system, underlying the importance of understanding the NC-cell interaction.

2. Materials and Methods

2.1. Experimental Setup (Optical Tweezers)

The OT system (Figure 1) forms two optical traps and allows for simultaneous manipulation of two RBCs. The traps are formed by a Nd:YAG infrared laser ILML3IF-300 (Leadlight Technology, Hsin-Chu, Taiwan) beam with output power up to 350 mW at a wavelength of 1064 nm.

The overall power of the initial beam is controlled by a half-wave plate $\lambda/2$ and a first beamsplitter cube; the power of the individual beams is regulated by a combination of $\lambda/2$ plates (not shown) and the second beamsplitter cube. The position of the second trap inside the sample in the focal plane of the focusing objective is adjusted by a conjugated beam tuneable mirror. After passing through the beam expander, light reaches the dichroic mirror, which reflects the laser beam to the sample. Tightly focusing of the laser beam was realized by an objective lens of high numerical aperture (NA = 1.0, water immersion LUMPlanFl, lOlympus, Tokyo, Japan), forming two traps inside the sample chamber. The sample illuminated with a white LED is transmitted through the infrared filter (IR-filter) to the complementary metal-oxide-semiconductor (CMOS) camera (Pixelink PL-B621M, lPixelink, Ottawa, Canada).

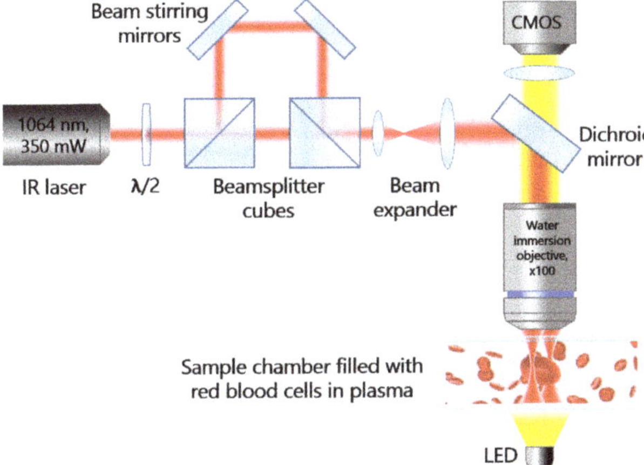

Figure 1. Schematic representation of the Optical tweezers (OT) setup. Optical traps are formed inside the sample chamber with a water immersion objective with high numerical aperture ($100\times$, $NA = 1$). Measurement procedure was imaged in the transmission mode and recorded with the complementary metal-oxide-semiconductor (CMOS) camera.

To calibrate the OT a viscous friction method was applied based on Stokes' law. The drag force is

$$F_d = 6\pi\eta RvK, \tag{1}$$

where η is the dynamic viscosity of blood plasma (1.2×10 mPa·s) [49], r is the RBC radius considering an RBC to be an equivolume sphere with r = 2.7 µm , v is the flow velocity relative to the object (velocity of the XY-stage movement), K is the correction factor for the ellipsoid.

Experiments were conducted at a 30-µm distance above the bottom of the cuvette to avoid interaction with other RBCs and bottom friction. The aggregation force was matched with the trapping force. The power at the moment when an RBC escapes the trap at a given velocity of the flow is equal to the trapping force.

2.2. Measurement Procedure

Measurements were performed as in our previous studies of mutual interaction of RBCs [50]. The aggregation force Fa of RBC is the minimum force required to stop two RBCs from overlapping. For a single measurement, two RBCs were trapped with individual optical traps, lifted for 30 µm and brought in contact at a small overlapping distance of $\Delta x \approx 2$ µm . From this stable position, the power of one trap starts slowly decreasing until spontaneous aggregation takes place and RBCs overlap. The power of the attenuated laser beam registered at this moment is equal to the aggregation force between the two RBCs. Disaggregation force (the force applied to separate two overlapped RBCs) was also measured to estimate energy density of RBC interaction.

2.3. Preparation of RBC Samples

Blood from the same donor was drawn by venepuncture directly in (EDTA)-covered tubes to prevent coagulation. The experiments were conducted in accordance with the obtained ethical permission from the Finnish Red Cross (No. 11/2019). In order to obtain platelet-free plasma, the whole blood was centrifuged at $3000 \times g$ for 10 min, plasma supernatant was aspirated gently through a micropipette and the procedure was repeated to remove the rest of the RBCs. Experiments were conducted at room temperature (20 °C).

RBCs were collected prior to experiments by finger pricking from a healthy donor (the same person was used as for plasma retrieval). A written consent was obtained from the volunteer. RBCs from the blood sample (hematocrit 45%) were washed with Dulbecco's Phosphate Buffered Saline (PBS, pH 7.4) by centrifugation at $4000 \times g$ for 10 min. Sedimented RBCs were accurately taken from the vial and suspended in blood plasma at a concentration of 0.5%.

NCs were dispersed in distilled water and placed in an ultrasonic bath for 5 min to destroy aggregates. Then NCs were added to the RBCs suspended in plasma following incubation for 1 h. The concentration of NCs was high enough to ensure interaction of RBCs with NCs—0.050106 per µL (with RBC concentration of 0.045×10^6 per µL). NCs were easily visible via the objective $100 \times$ during the experiments.

A sample chamber from a glass slide with a glass coverslip separated by a double-sided adhesive tape to form a 0.1-mm-thick sample gap was used in the measurements. Opened edges of the cuvette were isolated with Vaseline to prevent drying of the sample and the thermal drift. The plasma suspension of RBCs incubated with NCs was pipetted into the sample chamber to perform measurements.

2.4. Synthesis of Nanocapsules

The following chemicals obtained from Sigma-Aldrich were used to synthesize NCs: dextran sulfate, sodium salt (DS, MW > 70,000), poly-L-arginine hydrochloride (PARG, MW > 70,000), phosphate buffered saline (PBS, 0.01 M), calcium chloride dihydrate, ethylenediaminetetraacetic acid (EDTA) disodium salt, rhodamine ltetramethylrhodamine (TRITC) isothiocyanate (MW 536.08), anhydrous sodium carbonate, sodium chloride, citrate-stabilized magnetite nanoparticles (FeNPs), goat anti-Mouse IgG2a Secondary Antibody labeled by Alexa Fluor 546 (A-21133, Thermo Fisher Scientific, Waltham, MA, USA). Control siRNA labelled with Alexa 546 were purchased from Qiagen. Synthesis of Fe_3O_4 nanoparticles was performed using a home-built setup similar to that described by German et al. [15].

NCs were prepared using calcium carbonate ($CaCO_3$) particles as a sacrificial template. The $CaCO_3$ particles were fabricated as before [16]. $CaCl_2$ and Na_2CO_3 aqueous solutions (0.33 M) were mixed under vigorous stirring for 3 h and were dissolved in 10 mL ethylene glycol, leading to precipitation of $CaCO_3$ particles. When finished, $CaCO_3$ particles were washed with deionized water to remove unreacted species. The spherical $CaCO_3$ submicron particles with an average diameter of 400–600 nm were prepared. NCs were manufactured using the layer-by-layer (LbL) assembly technique by alternative deposition of oppositely charged polyelectrolytes on the $CaCO_3$ particles. For the layers, biocompatible polyelectrolytes such as dextran sulfate (DS) with the concentration of 1 mg/mL (2 mL) and poly-L-arginine hydrochloride (PARG) of 1 mg/mL (1 mL) were used. The average size of the core-shell capsules calculated from scanning electron microscopy (SEM) images was 640 ± 100 nm, shown on Figure 2a). The capsules were labelled by Rhodamine TRITC dye isothiocyanate and RNA molecules were adsorbed on the $CaCO_3$ surface. Encapsulation of the dye was performed during LbL. Fluorescence of the loaded Rhodamine TRITC dye was measured by confocal microscopy (see Figure 2b).

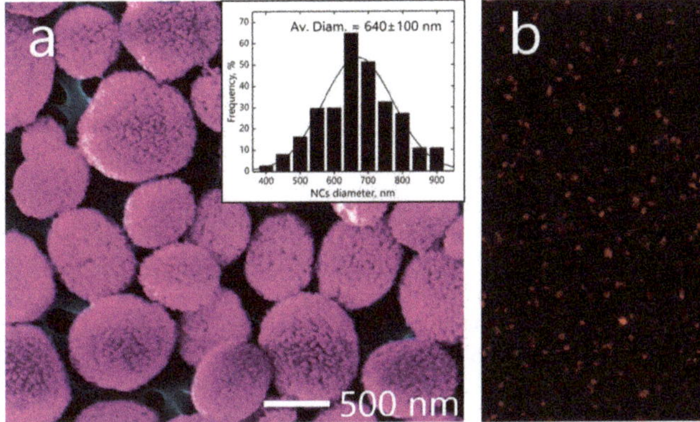

Figure 2. (a) Coloured scanning electron microscopy (SEM) images of core-shell polymeric nano-capsules (NCs) with size distribution with Gaussian fitting. (b) The red fluorescence on the confocal image comes from the shell layers containing rhodamine-ltetramethylrhodamine (TRITC).

Four types of polymeric core-shell NCs produced from the $CaCO_3$ particles, differed by the material embedded into the shells, were tested in the present study:

1. Secondary antibody NCs are NCs labeled with secondary antibody anti-mouse Alexa Fluor 546. NCs are often covered with antibodies to provide immuno-specific binding to the target cells [51].
2. RNA-NCs are NCs with RNA-labeled Rhodamine-TRITC, widely used for cancer theranostics, and can be applied for genome editing [52,53].
3. Fe_3O_4 magnetite NCs are NCs loaded with magnetic particles, which allow the magnetically assisted delivery, controlled drug release, and MRI imaging [54].
4. Rhodamine-labelled NCs are NCs labeled only with Rhodamine TRITC in a shell, which is used to control the allocation of NCs [16].

2.5. Preparation of RCBs for Optical and Scanning Electron Microscopy (SEM)

Washed RBCs were incubated with NCs in PBS at the concentration mentioned above within 1 h followed by their fixation with 1% glutaraldehyde (lMerck, Kenilworth, NJ, USA) for 15 min. The fixative solution was then removed from the vial and replaced with distilled water. A droplet of RBC suspension was pipetted onto a paper filter attached to a carbon tape. After evacuation the dried samples were coated with a 5-nm-thick platinum layer for electron microscopy (Zeiss Ultra Plus and Sigma FESEM, lCarl Zeiss, Oberkochen, Germany).

2.6. Estimation of the Human Mesenchymal Stem Cells (HMSCS) Morphological Properties

Human mesenchymal stem cells (hMSCs) were obtained from the bone marrow of healthy donors who gave their informed consent. Plating procedure was applied for cell isolation: 1 mL of heparinized bone marrow was resuspended in an α minimum essential medium (Lonza, Basel, Switzerland) supplemented with 100 IU/mL penicillin, 0.1 mg/mL streptomycin (lBiolot, Saint Petersburg, Russia), 10 vol % fetal bovine serum (FBS, HyClone, Logan, UT, USA), and 2 mM UltraGlutamine I (Lonza, Switzerland). The cells were cultured in Dulbecco's Modified Eagle Medium (DMEM) supplemented with 10% inactivated Fetal Bovine Serum (lThermo-Fischer Scientific, Waltham, MA, USA), under standard conditions (37 °C, 5% of CO_2, humidified sterile environment) to >85% confluency. For estimation of the cell morphological properties, the cells were incubated with NCs for 24 h; then they were fixed with 4% paraformaldehyde solution for immunostaining.

3. Results

Synthesized NCs differed only by the material loaded into the capsule's shell. Loaded material have no effect on a surface charge of NCs, since measured zeta potential value is 40 mV for all the NCs types. This value is associated with good or moderate stability of NCs in water solutions. The particular interest of the study is if the human blood plasma affecting NCs properties. OT measurements of RBC were performed in vitro in autologous blood plasma mimicking in vivo conditions.

After incubation of red blood cells (RBCs) with NCs in buffer solution, sample was diluted in blood plasma. NCs being transferred into blood plasma tend to form agglomerates, which are visible and easily distinguished in the sample (Figure 3a–d). Magnetite Fe_3O_4 and rhodamine-labelled NCs have formed the biggest aggregates among observed, exceeding the size of RBCs and reaching 10–15 μm, encircled on the Figure 3c,d. While NCs loaded with secondary antibody and RNA formed smaller aggregates and exhibited an adequate stability in blood plasma.

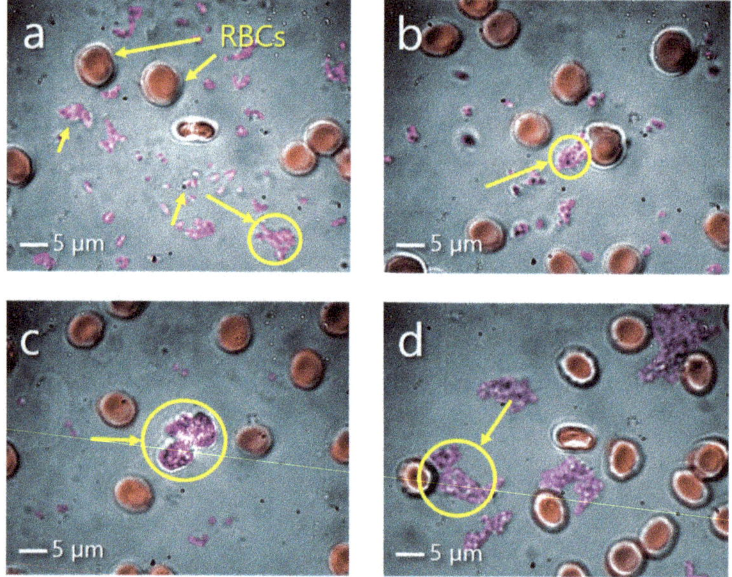

Figure 3. Optical microscopy images (objective 100×) of lred blood cells (RBCs) in blood plasma with NCs in false color (NC aggregates are encircled) during measurements with OT: (**a**) Secondary antibody NCs; (**b**) lribonucleic acid (RNA) NCs; (**c**) Fe_3O_4 NCs; (**d**) rhodamine-labelled NCs.

Aggregation forces between RBCs were measured on 20 pairs of cells for each sample. The measurements of the aggregation forces between RBCs showed that compared to the control measurements in the blood plasma without NCs treatment ($F_a = 4.26 \pm 1.87$ pN), no statistically significant differences (Mann-Whitney U-test at $p < 0.05$) were found after RBCs incubation with NCs, regardless the type of NCs. The aggregation force between RBCs was found to be almost the same for all samples (Figure 4a), $F_a = 5.38 \pm 1.56$ pN for the secondary antibody NCs, $F_a = 6.31 \pm 0.68$ pN for RNA NCs, $F_a = 5.87 \pm 1.08$ pN for Fe$_3$O$_4$ NCs, and $F_a = 4.78 \pm 1.01$ pN for Rhodamine-labelled NCs.

Interaction energy density between RBCs was calculated as described before [48]. The obtained dependence demonstrates the growth of the interaction energy between RBCs upon their separation with OT (Figure 4b). This dependence is typical for the so called "cross-bridges" model, assuming the accumulation of "cross-bridges" formed by plasma proteins between the interacting RBCs. RBCs interaction energy always increases upon cell disaggregation, as it was confirmed before in our studies [50,55,56]. Here, RBCs in presence of all the types of NCs demonstrate the same dependency, suggesting no difference in the interaction mechanism between the cells.

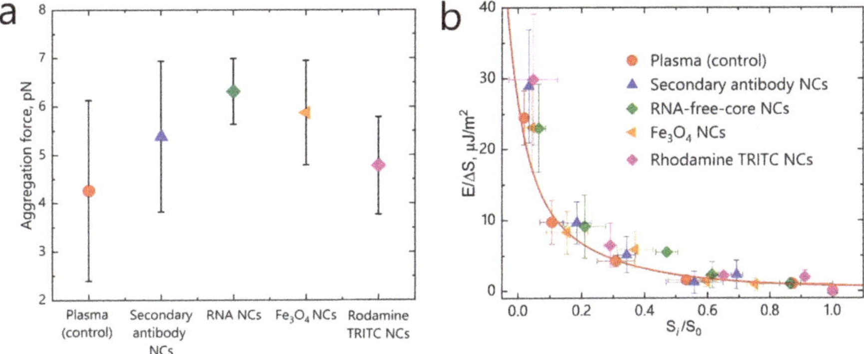

Figure 4. RBC aggregation force (**a**) and interaction energy density dependence on interaction area (**b**) between RBCs in blood plasma alone (control) and in presence of NCs measured with OT. ΔS is conjugated interaction area, S_i / S_0 is relative displacement of RBCs from the initial overlapping area S_0. The red curve corresponds to the cross-bridges model.

Analyzing SEM images of RBCs incubated with NCs (approximately 100s of cells), no visible adverse effects on the RBC morphology were found. NCs are too big to be able to penetrate RBCs and the only expected way of NCs influence on RBC is through the interaction/attachment of NCs to the cell membrane, possibly causing membrane deformations or ruptures. According to the SEM images RBCs preserved their native shape even with few NCs being found attached to a single cell (Figure 5b). Besides, no visible changes of the RBC shape were observed during OT measurements with 100× magnification.

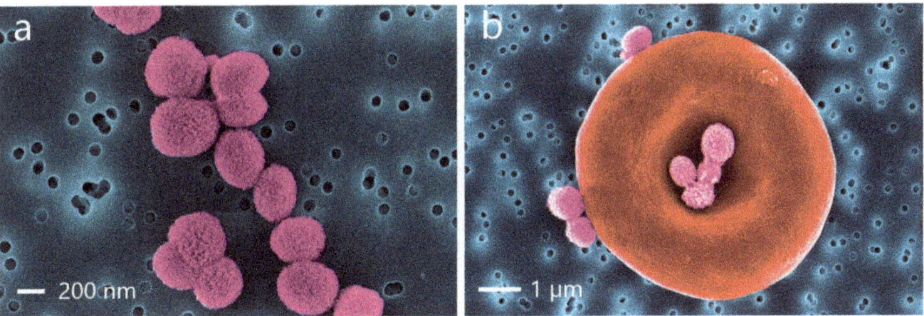

Figure 5. Coloured SEM images of (**a**) core-shell NCs with RNA; (**b**) rhodamine-labelled NCs attached to a RBC.

When NCs are incubated with hMSC, as seen in Figure 6a, changes in the cell structure were not observed. hMSCs are significantly larger cells of 20–30 μm, comparing to RBCs, and NCs can potentially penetrate the cell. Phase images (see Figure 6b) allow for evaluating the cell structure integrity. Capsules possess neutral pH of about 7, which is equal to the cell's pH, implying the biocompatibility of NCs. This observation proves the idea that NCs do not possess the adverse cytotoxic effect on the morphological properties of cells.

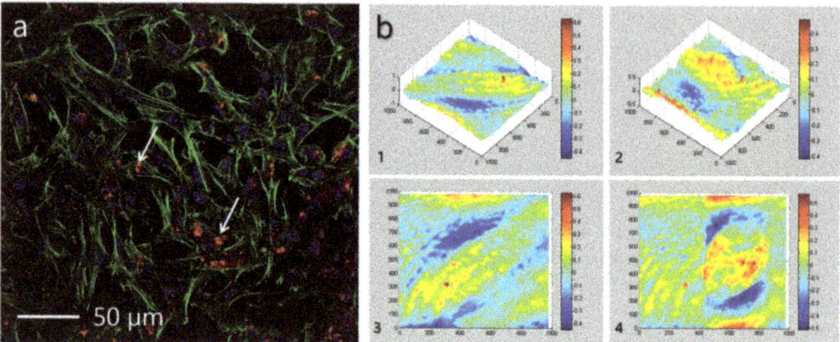

Figure 6. Interaction of polymeric NCs with Human Mesenchymal Stem Cells (hMSCs) elucidated with optical confocal microscopy. (**a**) Confocal image of hMSCa after 24 h incubation with NCs, where the green color denotes phalloidin lfluorescein (FITC) and the blue color denotes 4′,6-diamidino-2-phenylindole (DAPI) dyes. The arrows indicate fluorescing rhodamine-labelled NCs; (**b**) Phase images of hMSCs without capsules - initially: 3D (1), 2D (3) and after 24 h incubation with capsules: 3D (2), 2D (4).

4. Discussion

Since NCs carry positive charge, the preferable interaction with the negatively charged RBC membranes was expected. However, in blood plasma due to the presence of different proteins contributing to complex interaction of NCs with the surrounding media, NCs surface properties change. This results in the strong agglomeration observed for magnetite Fe_3O_4 and rhodamine-labelled NCs.

The tendency of NCs for agglomeration significantly decreases interaction between RBC membranes and NCs. In this way aggregation forces between the cells will be the same as in the control measurements. Interaction mechanism between RBCs have not changed for the cells incubated with NCs. Plasma protein adsorption can lead to the change of the NC surfaces properties, forming so-called "corona", which is able to decrease cytotoxic effects of the positively charged nanoparticles [57]. Thus, even if Fe_3O_4 and Rhodamine-labelled NCs do not cause RBC aggregation, strong NCs self-aggregation can reduce the efficiency of NCs injected into the blood stream and potentially cause blockage of small vessels. After incubation of negatively charged hMSCs with NCs, the latter were internalized inside the cell's matrices through the hMSC phagocytosis mechanism. Based on above, NCs charge can be considered as the defining mechanism for NC-cell interactions, due to the implications introduced by media (plasma) or cell intrinsic properties.

Incubation of RBCs with NCs in PBS resulted in the attachment of NCs to an RBC membrane, forming RBC-NCs complex, as it follows from the SEM. Probably, interaction between the cell and NCs is weak, so the complexes are falling apart and being transferred into blood plasma. For the desired coupling of RBCs with NCs proper NCs functionalization should be addressed. Regardless the material being embedded into the NCs shell, NCs show no influence of RBC interaction properties.

NCs must meet basic requirements for medical applications such as biocompatibility and ability to reach a target tissue. The described OT technology allows for fast evaluation of NC influence on the aggregation state of individual RBCs. The obtained information enables finding safe concentrations of NCs to prevent their adverse influence on blood rheological properties, to reveal the influence o material loaded into the NCs on cell state, decreasing in this way the risk of side effects.

For the integral assessment of NCs, other blood components should be also involved, such as platelets and leukocytes. OT and microscopy based approaches conjugated with microfluidic platforms, can be transformed into the blood vessel modelling system, to reveal the effects of different nanomaterials on cells, minimizing the use of animals for testing.

The use of fluorescence spectroscopy, in combination with phase microscopy, can also increase the statistical significance and reliability of clinical trials and reduce the number of animals needed and decrease their suffering. The data obtained by this method provides information about the pharmacodynamics and the optimal dosage of the drug. To carry out clinical trials, it is necessary to estimate the capsules influence on the morphology of cells. HMSCs after long incubation with NCs preserved their integrity and viability, despite the evenly distributed NCs inside the cells.

5. Conclusions

OT and optical microscopy observations have demonstrated different behavior of NCs in blood plasma. The magnetite and rhodamine-labelled NCs are prone to strong agglomeration in plasma, while secondary antibody and RNA-NCs are relatively stable and uniformly distributed. NCs self-aggregation can potentially result in a microvascular blockage of small blood capillaries in vivo. None of the tested NCs caused changes of aggregation forces of RBCs compared to the control measurements. In all cases, the cross-bridges model of RBC interaction is applicable, confirming the same interaction mechanism between RBCs in presence of NCs. Combination of OT and SEM imaging together with conventional optical microscopy shows that none of the types of NCs used have any adverse influence on RBCs' morphology, or on hMSCs, indicating that polymeric NCs are not cytotoxic for these cells in the tested conditions.

Author Contributions: Conceptualization, I.M., Y.T., T.A., I.S., S.V.; Methodology, Y.T., T.A., A.P.; Investigation, T.A., Y.T. and R.Z.; Funding acquisition, S.V., and I.M.; Writing—original draft preparation Y.T., T.A. and I.M.; Writing—review and editing, T.A., I.M., and A.P.; Supervision, I.M., A.B., A.P., S.V. and I.S. All authors have read and agreed to the published version of the manuscript.

Funding: EDUFI Fellowship (TM-17-10370, TM-18-10820, T.A.), Oulun yliopiston apuraharahasto, No. 20180107, Suomen Kulttuurirahasto (grant No. 00190188, T.A.), CIMO Fellowship (TM-15-9729, Y.T.), EDUFI Fellowship (TM-17-10389, Y.T.), Academy of Finland (project no. 326126, Y.T.), China Scholarship Council (CSC No. 201706410089, R.Z.), Tauno Tönning Foundation (grant No. 20190104, R.Z.), Academy of Finland (grant No. 290596, 314369, 326126, A.B., A.P.). The authors acknowledge contribution of Russian Science Foundation (Project 19-72-30012) for the fabrication of some samples. I.M. also acknowledges partial support from the Academy of Finland (project 326204, 325097), MEPhI Academic Excellence Project (Contract No. 02.a03.21.0005), and the National Research Tomsk State University Academic D.I. Mendeleev Fund Program.

Acknowledgments: Authors acknowledge the initial involvement of M. Kinnunen (University of Oulu, Finland), A.V. Priezzhev (M.V. Lomonosov Moscow State University, Russia) and A. Karmenyan (Physics Department of National Don Hwa University, Hualien, Taiwan) at the early stage of the OT development.

Conflicts of Interest: The authors declare no conflict of interest.

Abbreviations

The following abbreviations are used in this manuscript:

OT	Optical tweezers
NCs	Nanocapsules
RBCs	Red blood cells
hMSCs	human mesenchymal stem cells

References

1. Yoo, J.W.; Irvine, D.J.; Discher, D.E.; Mitragotri, S. Bio-inspired, bioengineered and biomimetic drug delivery carriers. *Nat. Rev. Drug Discov.* **2011**, *10*, 521. [CrossRef]
2. Lockney, D.; Franzen, S.; Lommel, S. Viruses as nanomaterials for drug delivery. In *Biomedical Nanotechnology*; Springer: Berlin/Heidelberg, Germany, 2011; pp. 207–221.

3. Kolesnikova, T.A.; Skirtach, A.G.; Moehwald, H. Red blood cells and polyelectrolyte multilayer capsules: natural carriers versus polymer-based drug delivery vehicles. *Expert Opin. Drug Deliv.* **2013**, *10*, 47–58. [CrossRef]

4. Magnani, M.; Chiarantini, L.; Vittoria, E.; Mancini, U.; Rossi, L.; Fazi, A. Red blood cells as an antigen-delivery system. *Biotechnol. Appl. Biochem.* **1992**, *16*, 188–194.

5. Muzykantov, V.R. Drug delivery by red blood cells: Vascular carriers designed by mother nature. *Expert Opin. Drug Deliv.* **2010**, *7*, 403–427. [CrossRef]

6. Deville, S.; Hadiwikarta, W.W.; Smisdom, N.; Wathiong, B.; Ameloot, M.; Nelissen, I.; Hooyberghs, J. Transient loading of CD34+ hematopoietic progenitor cells with polystyrene nanoparticles. *Int. J. Nanomed.* **2017**, *12*, 459. [CrossRef]

7. Peer, D. Immunotoxicity derived from manipulating leukocytes with lipid-based nanoparticles. *Adv. Drug Deliv. Rev.* **2012**, *64*, 1738–1748. [CrossRef]

8. Textor, J.A.; Clark, K.C.; Walker, N.J.; Aristizobal, F.A.; Kol, A.; LeJeune, S.S.; Bledsoe, A.; Davidyan, A.; Gray, S.N.; Bohannon-Worsley, L.K.; et al. Allogeneic stem cells alter gene expression and improve healing of distal limb wounds in horses. *Stem Cells Transl. Med.* **2018**, *7*, 98–108. [CrossRef]

9. Li, M.; Guo, K.; Ikehara, S. Stem cell treatment for Alzheimer's disease. *Int. J. Mol. Sci.* **2014**, *15*, 19226–19238. [CrossRef]

10. Ansari, S.; Diniz, I.M.; Chen, C.; Aghaloo, T.; Wu, B.M.; Shi, S.; Moshaverinia, A. Alginate/hyaluronic acid hydrogel delivery system characteristics regulate the differentiation of periodontal ligament stem cells toward chondrogenic lineage. *J. Mater. Sci. Mater. Med.* **2017**, *28*, 162. [CrossRef]

11. Perteghella, S.; Martella, E.; De Girolamo, L.; Perucca Orfei, C.; Pierini, M.; Fumagalli, V.; Pintacuda, D.; Chlapanidas, T.; Viganò, M.; Faragò, S.; et al. Fabrication of innovative silk/alginate microcarriers for mesenchymal stem cell delivery and tissue regeneration. *Int. J. Mol. Sci.* **2017**, *18*, 1829. [CrossRef]

12. Bochot, A.; Fattal, E. Liposomes for intravitreal drug delivery: A state of the art. *J. Control. Release* **2012**, *161*, 628–634. [CrossRef]

13. Gaucher, G.; Satturwar, P.; Jones, M.C.; Furtos, A.; Leroux, J.C. Polymeric micelles for oral drug delivery. *Eur. J. Pharm. Biopharm.* **2010**, *76*, 147–158. [CrossRef]

14. Parakhonskiy, B.V.; Yashchenok, A.M.; M'ohwald, H.; Volodkin, D.; Skirtach, A.G. Release from polyelectrolyte multilayer capsules in solution and on polymeric surfaces. *Adv. Mater. Interfaces* **2017**, *4*, 1600273. [CrossRef]

15. German, S.; Inozemtseva, O.; Navolokin, N.; Pudovkina, E.; Zuev, V.; Volkova, E.; Bucharskaya, A.; Pleskova, S.; Maslyakova, G.; Gorin, D. Synthesis of magnetite hydrosols and assessment of their impact on living systems at the cellular and tissue levels using MRI and morphological investigation. *Nanotechnol. Russ.* **2013**, *8*, 573–580. [CrossRef]

16. Tarakanchikova, Y.; Stelmashchuk, O.; Seryogina, E.; Piavchenko, G.; Zherebtsov, E.; Dunaev, A.; Popov, A.; Meglinski, I. Allocation of rhodamine-loaded nanocapsules from blood circulatory system to adjacent tissues assessed in vivo by fluorescence spectroscopy. *Laser Phys. Lett.* **2018**, *15*, 105601. [CrossRef]

17. Svenskaya, Y.I.; Fattah, H.; Zakharevich, A.; Gorin, D.A.; Sukhorukov, G.B.; Parakhonskiy, B. Ultrasonically assisted fabrication of vaterite submicron-sized carriers. *Adv. Powder Technol.* **2016**, *27*, 618–624. [CrossRef]

18. Rogers, T.L.; Wallick, D. Reviewing the use of ethylcellulose, methylcellulose and hypromellose in microencapsulation. Part 1: materials used to formulate microcapsules. *Drug Dev. Ind. Pharm.* **2012**, *38*, 129–157. [CrossRef]

19. Saveleva, M.S.; Lengert, E.V.; Gorin, D.A.; Parakhonskiy, B.V.; Skirtach, A.G. Polymeric and lipid membranes—From spheres to flat membranes and vice versa. *Membranes* **2017**, *7*, 44. [CrossRef]

20. Quadir, M.A.; Haag, R. Biofunctional nanosystems based on dendritic polymers. *J. Control. Release* **2012**, *161*, 484–495. [CrossRef]

21. Lengert, E.; Saveleva, M.; Abalymov, A.; Atkin, V.; Wuytens, P.C.; Kamyshinsky, R.; Vasiliev, A.L.; Gorin, D.A.; Sukhorukov, G.B.; Skirtach, A.G.; et al. Silver alginate hydrogel micro-and nanocontainers for theranostics: synthesis, encapsulation, remote release, and detection. *ACS Appl. Mater. Interfaces* **2017**, *9*, 21949–21958. [CrossRef]

22. Prokop, A.; Davidson, J.M. Nanovehicular intracellular delivery systems. *J. Pharm. Sci.* **2008**, *97*, 3518–3590. [CrossRef] [PubMed]

23. Ai, H.; Jones, S.A.; Lvov, Y.M. Biomedical applications of electrostatic layer-by-layer nano-assembly of polymers, enzymes, and nanoparticles. *Cell Biochem. Biophys.* **2003**, *39*, 23. [CrossRef]
24. Tong, W.; Song, X.; Gao, C. Layer-by-layer assembly of microcapsules and their biomedical applications. *Chem. Soc. Rev.* **2012**, *41*, 6103–6124. [CrossRef]
25. De Koker, S.; De Cock, L.J.; Rivera-Gil, P.; Parak, W.J.; Velty, R.A.; Vervaet, C.; Remon, J.P.; Grooten, J.; De Geest, B.G. Polymeric multilayer capsules delivering biotherapeutics. *Adv. Drug Deliv. Rev.* **2011**, *63*, 748–761. [CrossRef]
26. Richardson, J.J.; Choy, M.Y.; Guo, J.; Liang, K.; Alt, K.; Ping, Y.; Cui, J.; Law, L.S.; Hagemeyer, C.E.; Caruso, F. Polymer Capsules for Plaque-Targeted In Vivo Delivery. *Adv. Mater.* **2016**, *28*, 7703–7707. [CrossRef]
27. Arruebo, M. Drug delivery from structured porous inorganic materials. *Wiley Interdiscip. Rev. Nanomed. Nanobiotechnol.* **2012**, *4*, 16–30. [CrossRef]
28. Svenskaya, Y.; Parakhonskiy, B.; Haase, A.; Atkin, V.; Lukyanets, E.; Gorin, D.; Antolini, R. Anticancer drug delivery system based on calcium carbonate particles loaded with a photosensitizer. *Biophys. Chem.* **2013**, *182*, 11–15. [CrossRef]
29. Kilic, E.; Novoselova, M.V.; Lim, S.H.; Pyataev, N.A.; Pinyaev, S.I.; Kulikov, O.A.; Sindeeva, O.A.; Mayorova, O.A.; Murney, R.; Antipina, M.N.; et al. Formulation for oral delivery of lactoferrin based on bovine serum albumin and tannic acid multilayer microcapsules. *Sci. Rep.* **2017**, *7*, 44159. [CrossRef]
30. Varela-Eirin, M.; Varela-Vazquez, A.; Mateos, M.R.C.; Vila-Sanjurjo, A.; Fonseca, E.; Mascareñas, J.L.; Vázquez, M.E.; Mayan, M.D. Recruitment of RNA molecules by connexin RNA-binding motifs: Implication in RNA and DNA transport through microvesicles and exosomes. *Biochim. Biophys. Acta BBA Mol. Cell Res.* **2017**, *1864*, 728–736. [CrossRef]
31. JohnáTng, W.; HuiáNg, H. Layered polymeric capsules inhibiting the activity of RNases for intracellular delivery of messenger RNA. *J. Mater. Chem.* **2015**, *3*, 5842–5848.
32. Timin, A.S.; Muslimov, A.R.; Petrova, A.V.; Lepik, K.V.; Okilova, M.V.; Vasin, A.V.; Afanasyev, B.V.; Sukhorukov, G.B. Hybrid inorganic-organic capsules for efficient intracellular delivery of novel siRNAs against influenza A (H1N1) virus infection. *Sci. Rep.* **2017**, *7*, 102. [CrossRef]
33. Chen, X.; Cui, J.; Ping, Y.; Suma, T.; Cavalieri, F.; Besford, Q.A.; Chen, G.; Braunger, J.A.; Caruso, F. Probing cell internalisation mechanics with polymer capsules. *Nanoscale* **2016**, *8*, 17096–17101. [CrossRef]
34. Sun, H.; Cui, J.; Ju, Y.; Chen, X.; Wong, E.H.; Tran, J.; Qiao, G.G.; Caruso, F. Tuning the properties of polymer capsules for cellular interactions. *Bioconj. Chem.* **2017**, *28*, 1859–1866. [CrossRef]
35. Timin, A.S.; Muslimov, A.R.; Lepik, K.V.; Saprykina, N.N.; Sergeev, V.S.; Afanasyev, B.V.; Vilesov, A.D.; Sukhorukov, G.B. Triple-responsive inorganic–organic hybrid microcapsules as a biocompatible smart platform for the delivery of small molecules. *J. Mater. Chem.* **2016**, *4*, 7270–7282. [CrossRef]
36. Bharali, D.J.; Sahoo, S.K.; Mozumdar, S.; Maitra, A. Cross-linked polyvinylpyrrolidone nanoparticles: A potential carrier for hydrophilic drugs. *J. Colloid Interface Sci.* **2003**, *258*, 415–423. [CrossRef]
37. Bharali, D.J.; Klejbor, I.; Stachowiak, E.K.; Dutta, P.; Roy, I.; Kaur, N.; Bergey, E.J.; Prasad, P.N.; Stachowiak, M.K. Organically modified silica nanoparticles: a nonviral vector for in vivo gene delivery and expression in the brain. *Proc. Natl. Acad. Sci. USA* **2005**, *102*, 11539–11544. [CrossRef]
38. Rejeeth, C.; Kannan, S.; Muthuchelian, K. Development of in vitro gene delivery system using ORMOSIL nanoparticle: Analysis of p53 gene expression in cultured breast cancer cell (MCF-7). *Cancer Nanotechnol.* **2012**, *3*, 55. [CrossRef]
39. Chambers, E.; Mitragotri, S. Long circulating nanoparticles via adhesion on red blood cells: Mechanism and extended circulation. *Exp. Biol. Med.* **2007**, *232*, 958–966.
40. Rothen-Rutishauser, B.M.; Schurch, S.; Haenni, B.; Kapp, N.; Gehr, P. Interaction of fine particles and nanoparticles with red blood cells visualized with advanced microscopic techniques. *Environ. Sci. Technol.* **2006**, *40*, 4353–4359. [CrossRef]
41. Villa, C.H.; Anselmo, A.C.; Mitragotri, S.; Muzykantov, V. Red blood cells: Supercarriers for drugs, biologicals, and nanoparticles and inspiration for advanced delivery systems. *Adv. Drug Deliv. Rev.* **2016**, *106*, 88–103. [CrossRef]
42. Timin, A.S.; Peltek, O.O.; Zyuzin, M.V.; Muslimov, A.R.; Karpov, T.E.; Epifanovskaya, O.S.; Shakirova, A.I.; Zhukov, M.V.; Tarakanchikova, Y.V.; Lepik, K.V.; et al. Safe and Effective Delivery of Antitumor Drug Using Mesenchymal Stem Cells Impregnated with Submicron Carriers. *ACS Appl. Mater. Interfaces* **2019**, *11*, 13091–13104. [CrossRef]

43. Pan, D.; Vargas-Morales, O.; Zern, B.; Anselmo, A.C.; Gupta, V.; Zakrewsky, M.; Mitragotri, S.; Muzykantov, V. The effect of polymeric nanoparticles on biocompatibility of carrier red blood cells. *PLoS ONE* **2016**, *11*, e0152074. [CrossRef]

44. Anselmo, A.C.; Mitragotri, S. Cell-mediated delivery of nanoparticles: Taking advantage of circulatory cells to target nanoparticles. *J. Control. Release* **2014**, *190*, 531–541. [CrossRef]

45. Roney, C.; Kulkarni, P.; Arora, V.; Antich, P.; Bonte, F.; Wu, A.; Mallikarjuana, N.; Manohar, S.; Liang, H.F.; Kulkarni, A.R.; et al. Targeted nanoparticles for drug delivery through the blood–brain barrier for Alzheimer's disease. *J. Control. Release* **2005**, *108*, 193–214. [CrossRef]

46. Ashkin, A. Acceleration and trapping of particles by radiation pressure. *Phys. Rev. Lett.* **1970**, *24*, 156. [CrossRef]

47. Ashkin, A.; Dziedzic, J.M.; Bjorkholm, J.; Chu, S. Observation of a single-beam gradient force optical trap for dielectric particles. *Opt. Lett.* **1986**, *11*, 288–290. [CrossRef]

48. Lee, K.; Kinnunen, M.; Khokhlova, M.D.; Lyubin, E.V.; Priezzhev, A.V.; Meglinski, I.; Fedyanin, A.A. Optical tweezers study of red blood cell aggregation and disaggregation in plasma and protein solutions. *J. Biomed. Opt.* **2016**, *21*, 035001. [CrossRef]

49. Késmárky, G.; Kenyeres, P.; Rábai, M.; Tóth, K. Plasma viscosity: A forgotten variable. *Clin. Hemorheol. Microcirc.* **2008**, *39*, 243–246. [CrossRef]

50. Avsievich, T.; Popov, A.; Bykov, A.; Meglinski, I. Mutual interaction of red blood cells assessed by optical tweezers and scanning electron microscopy imaging. *Opt. Lett.* **2018**, *43*, 3921–3924. [CrossRef]

51. Hervella, P.; Alonso-Sande, M.; Ledo, F.; L Lucero, M.; J Alonso, M.; Garcia-Fuentes, M. PEGylated lipid nanocapsules with improved drug encapsulation and controlled release properties. *Curr. Top. Med. Chem.* **2014**, *14*, 1115–1123. [CrossRef]

52. Zhu, G.; Mei, L.; Vishwasrao, H.D.; Jacobson, O.; Wang, Z.; Liu, Y.; Yung, B.C.; Fu, X.; Jin, A.; Niu, G.; et al. Intertwining DNA-RNA nanocapsules loaded with tumor neoantigens as synergistic nanovaccines for cancer immunotherapy. *Nat. Commun.* **2017**, *8*, 1482. [CrossRef]

53. Yan, M.; Wen, J.; Liang, M.; Lu, Y.; Kamata, M.; Chen, I.S. Modulation of gene expression by polymer nanocapsule delivery of DNA cassettes encoding small RNAs. *PLoS ONE* **2015**, *10*, e0127986. [CrossRef]

54. Chen, H.; Sulejmanovic, D.; Moore, T.; Colvin, D.C.; Qi, B.; Mefford, O.T.; Gore, J.C.; Alexis, F.; Hwu, S.J.; Anker, J.N. Iron-loaded magnetic nanocapsules for pH-triggered drug release and MRI imaging. *Chem. Mater.* **2014**, *26*, 2105–2112. [CrossRef]

55. Avsievich, T.; Popov, A.; Bykov, A.; Meglinski, I. Mutual interaction of red blood cells influenced by nanoparticles. *Sci. Rep.* **2019**, *9*, 5147. [CrossRef]

56. Zhu, R.; Avsievich, T.; Bykov, A.; Popov, A.; Meglinski, I. Influence of pulsed He–Ne Laser irradiation on the red blood cell interaction studied by optical tweezers. *Micromachines* **2019**, *10*, 853. [CrossRef]

57. Corbo, C.; Molinaro, R.; Parodi, A.; Toledano Furman, N.E.; Salvatore, F.; Tasciotti, E. The impact of nanoparticle protein corona on cytotoxicity, immunotoxicity and target drug delivery. *Nanomedicine* **2016**, *11*, 81–100. [CrossRef]

 *micromachines*

Article

Fusing Artificial Cell Compartments and Lipid Domains Using Optical Traps: A Tool to Modulate Membrane Composition and Phase Behaviour

Adithya Vivek [1], Guido Bolognesi [2] and Yuval Elani [3,4,*]

[1] Department of Chemistry, Imperial College London, London W12 0BZ, UK; adithya.vivek17@imperial.ac.uk
[2] Department of Chemical Engineering, Loughborough University, Leicestershire LE11 3TU, UK; G.Bolognesi@lboro.ac.uk
[3] Department of Chemical Engineering, Imperial College London, London SW7 2AZ, UK
[4] fabriCELL, Imperial College London, London SW7 2AZ, UK
* Correspondence: y.elani@imperial.ac.uk

Received: 6 January 2020; Accepted: 28 March 2020; Published: 7 April 2020

Abstract: New technologies for manipulating biomembranes have vast potential to aid the understanding of biological phenomena, and as tools to sculpt novel artificial cell architectures for synthetic biology. The manipulation and fusion of vesicles using optical traps is amongst the most promising due to the level of spatiotemporal control it affords. Herein, we conduct a suite of feasibility studies to show the potential of optical trapping technologies to (i) modulate the lipid composition of a vesicle by delivering new membrane material through fusion events and (ii) manipulate and controllably fuse coexisting membrane domains for the first time. We also outline some noteworthy morphologies and transitions that the vesicle undergoes during fusion, which gives us insight into the mechanisms at play. These results will guide future exploitation of laser-assisted membrane manipulation methods and feed into a technology roadmap for this emerging technology.

Keywords: optical traps; vesicles; artificial cells; membrane biophysics; phase separation; membranes

1. Introduction

Cell membranes are universal biological motifs that define cellular boundaries. They allow for cells to compartmentalise content, control the influx/efflux of materials, and they act as a surface on which biochemical reactions occur. Synthetic versions of cell membranes are not only used to shed light on fundamental principles underlying the function of membranes away from the complex cellular environment, but increasingly as components of artificial cells in the field of bottom-up synthetic biology [1–3].

The fusion of two cell membranes is a ubiquitous event in biological systems. Cell fusion is integral to a host of cellular processes, including endocytosis, exocytosis, cell division, cell-cell communication, and neurotransmission [4]. Bioengineers are increasingly exploiting this phenomenon, with synthetic membrane fusion events playing key roles in liposomal drug delivery [5], cell-like microreactors [6,7], cell transfection [8], the creation of cell hybrids for vaccine generation [9], and for therapeutic applications [10,11]. Membrane fusion is also being deployed in synthetic biology, to initiate protein synthesis in artificial cells through the delivery of genetic material [12] and in the creation of artificial cells that are capable of growing over time by subsuming new membrane material [13].

Fusion techniques generally fall into two camps. The first is bulk fusion, where the inherent biophysical properties of the membranes lead to stochastic fusion events in a vesicle population (e.g., through charge [14], membrane-bound DNA nanotechnology [15], and membrane mechanical properties [16]). The second is externally triggered fusion, where defined vesicles are brought into contact and fusion is initiated through applied forces (e.g., using electric fields [17,18] or lasers [12,19]). The attraction of the latter is the full level of spatiotemporal control of the fusion process that is available. Not only can a user define which vesicle will fuse with which, by the timing, spatial arrangement, and the number of fusion events can also be controlled.

There have been several recent studies on techniques to fuse Giant Unilamellar Vesicles (GUVs; enclosed cell-sized bilayer spheres) while using optical traps and membrane-bound gold nanoparticles (AuNP) [12,19,20]. The underlying mechanism behind this process is AuNP absorbance at the laser focus, followed by heat dissipation, local temperature elevation (>100 °C) in a region of characteristic size that is equal to the NP diameter [21], membrane expansion, and then fusion [19,20]. Similar optical tweezing technologies have recently been used to controllably manipulate co-existing membrane domains (analogues of cellular lipid rafts) within the membrane [22].

Several applications of laser-assisted vesicle fusion technologies have been demonstrated, including for the initiation of chemical reactions in vesicle microreactors [12], the creation of hybrid cells [20], and sequential dilution of encapsulated aqueous material [12]. However, the full capabilities of this technology have not been fully explored, and the potential for it to be used to modulate the composition and spatial arrangements of the membrane itself (and not simply mixing the encapsulated cargo) is largely an unchartered frontier.

Herein, we address this shortcoming through a series of proof-of-principle experiments on GUVs. Firstly, we show for the first time that laser-mediated vesicle fusion can be used to dynamically modulate the membrane composition and phase state by introducing new membrane material. Secondly, we use this technology to controllably fuse individual coexisting membrane domains with one another. Finally, we present a phenomenological account of some striking vesicle morphologies that the vesicle adopts during fusion, which have the potential to inform the underlying biophysical rules governing the fusion process. This work acts a roadmap for future work by highlighting the potential of optical trapping in membrane manipulation for a range of soft-matter biotechnology applications.

2. Results and Discussion

2.1. Dynamic Modulation of Lipid Composition through Vesicle Fusion

The use of optical traps to select, manipulate, and fuse user-defined vesicles with another opens the possibility of dynamically and controllably modulating membrane composition by introducing new lipid cargo. Although fusion between Small Unilamellar Vesicles (LUVs; c. 100 nm diameter) and GUVs has been used to introduce protein machinery into membranes [23] and induce GUV growth [13], this method cannot be used to controllably alter the lipid composition due to the lack of control of the number of fusion events. In principle, laser-assisted GUV fusion offers precise spatial and temporal control which bypasses this limitation. The experimental setup (Figure 1) consists of a single-beam optical trap that is built into an inverted epi-fluorescence microscope system allowing for the user to select AuNP labelled vesicles, bringing them into contact to form an adhesion patch (laser: 20–100 mW at trap), and induce fusion by increasing the laser power (>150 mW at trap). The adhesion patch is formed due to the presence of 0.2 mM NaCl in the external solution which screens out electrostatic repulsion between the vesicles, allowing attractive forces to dominate [12]. The fusion event is triggered by the interaction of the continuous laser beam with the nanoparticles sitting in the adhesion patch.

What enabled the manipulation of the lipid vesicles using an optical trap was the difference in molecular composition and, hence, refractive index between the vesicle interior (0.75 M sucrose) and exterior (0.35 M glucose, 0.2 M NaCl). This refractive index mismatch was enough to facilitate trapping and manipulation at laser powers above 20 mW at the trap.

We track changes in lipid composition post-fusion through the phase state of the membrane. Membranes can exist in a fully liquid phase (Lα), gel phase (Lβ) or have co-existing phase-separated domains [24]. In our experiments, there are either gel/liquid coexisting domains, or liquid-ordered/liquid-disordered (L_o/L_d) coexisting domains. The phase state is dependent on the precise lipid composition of the membrane [24,25], and it can be visualised using fluorescence microscopy through the incorporation of fluorescently labelled lipids (Rh-PE), which preferentially partitions into the disordered phases.

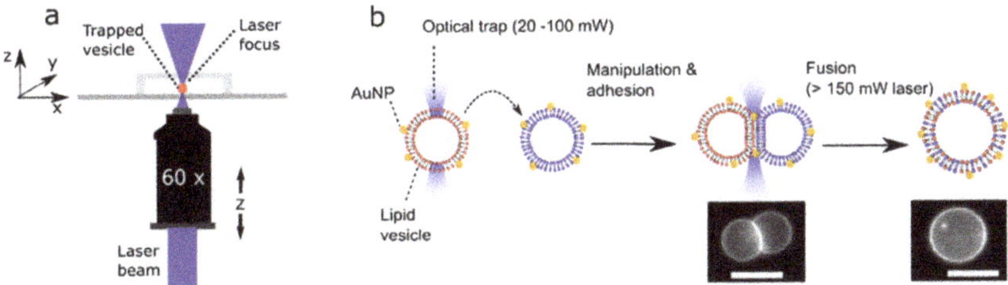

Figure 1. Schematic of the processes involved in laser-assisted vesicle adhesion and fusion. (**a**) Experimental setup. By moving a motorized stage and the objective itself, a vesicle could be manipulated in the x, y, and z direction relative to its surroundings. (**b**) Gold nanoparticles (AuNP) labelled vesicles of defined composition are brought together using an optical trap where they adhere due to the presence of NaCl in the external solution. Focussing the laser at the adhesion patch is used to initiate fusion, forming a unified structure with lipids originating from two previously distinct vesicles. Images of fluorescently labelled Giant Unilamellar Vesicles (GUVs) pre- and post-fusion, as shown. Scale bar = 5 μm.

In these experiments, we use ternary lipid mixtures. The lipids we used were: DOPC, which promotes the formation of a liquid phase (L_α) at room temperature; EggSM, which promotes the formation of a gel phase (L_β) at room temperature; and cholesterol, which promotes the formation of the liquid-ordered (L_o) phase in ternary mixtures. Using the laser-assisted fusion technique, we were able to move a vesicle from a gel/liquid coexisting regime to an L_o/L_d one, by delivering cholesterol-rich membrane material (Figure 2). The ternary lipid composition used for this experiment was DOPC/EggSM/Chol with the constituent lipids that were present in different ratios. Using iso-osmotic conditions, we fused two vesicles of c. 5 μm diameter, one of which was a 1:1:0 GUV (gel/liquid; small speckled non-spherical domains) with a 1:1:3 GUV (L_α; no domains, uniform fluorescence). This changed the membrane composition to lie in the L_o/L_d region of the phase diagram (large spherical domains) [24]. Assuming that the vesicles were identical in size, the final vesicle is expected to have a DOPC/EggSM/Chol 2:2:3 composition.

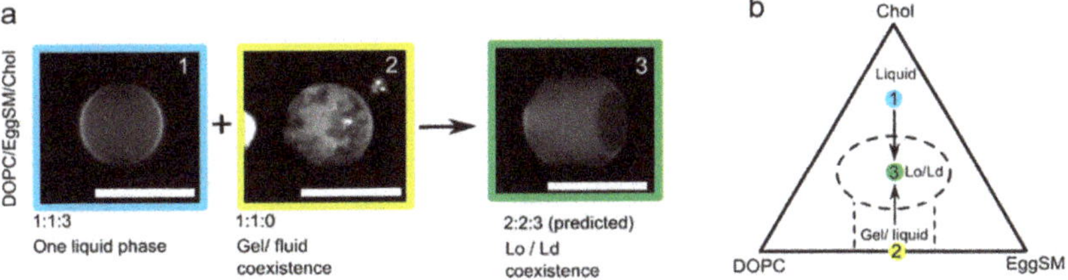

Figure 2. Changing membrane phase state through delivery of new lipid material. (**a**) Two giant vesicles of different compositions and phase states are fused together using an optical trap. Adding a fully fluid vesicle to one showing gel/fluid domain coexistence (irregularly shaped domains) yielded a vesicle exhibiting liquid-ordered/liquid disordered domain coexistence (spherical domains). Scale bar = 5 μm. (**b**) A schematic demonstrating adding lipid material from vesicle (1) to vesicle (2) leading to a new vesicle (3) occupying a different part of the phase diagram. Note: this is not an empirical phase diagram; it is an approximate one used for illustrative purposes only.

2.2. Manipulation and Fusion of Defined Coexisting Domains

In addition to manoeuvring and fusing membrane compartments with each other, we demonstrate that individual coexisting domains within a lipid membrane can be merged with spatiotemporal control. The ability to manipulate domains using optical forces has previously been shown [22] but using an optical trap to controllably fuse domains together has not. In a previous study [22], we have shown how the difference in both membrane refractive index and thickness between the L_o and L_d phase results in an optical gradient force that enables domain trapping and manipulation via a single trap. An approximated expression of the optical trapping force that was exerted by the laser on a trapped domain was also derived [22]. In the experiments below, we show that we can drag L_d domains across the membrane surface, and bring them in contact with adjacent domains to initiate fusion, by taking a DOPC/EggSM/Chol 1:1:1 vesicle and focussing the laser (0.23 W at trap) at the L_o/L_d interface (Figure 3; Video S1). We were able to sequentially fuse nine isolated domains to yield a domain of an incrementally increasing size using this method. We note that, with this composition, domains did not naturally coalesce within the timescales of the experiment, due them having a 'bulged' morphology, where they protrude out the vesicle body. Domain bulging means that the total interface length of the domain boundary is reduced, in turn reducing the total line tension, which is energetically favourable [24,26,27]. Ordinarily, domains would collide and coalesce, eventually growing into one large domain to reduce the total interface length. However, the bulged geometry creates repulsion between domains, allowing for multiple domains to be stabilized in a kinetically trapped regime. In our experiments, the repulsion between domains is overcome while using the optical trap through two processes operating in tandem. The first is that the traps are used to bring the domains together, overcoming the inter-domain repulsion. The second is through the heat emanating from the laser focus and the resulting temperature increase (maximum increase estimated to be 4–6 °C while using the laser powers used) [12]. This reduces the line tension, which in turn reduces the extent of domain bulging and the lowering of the inter-domain repulsion.

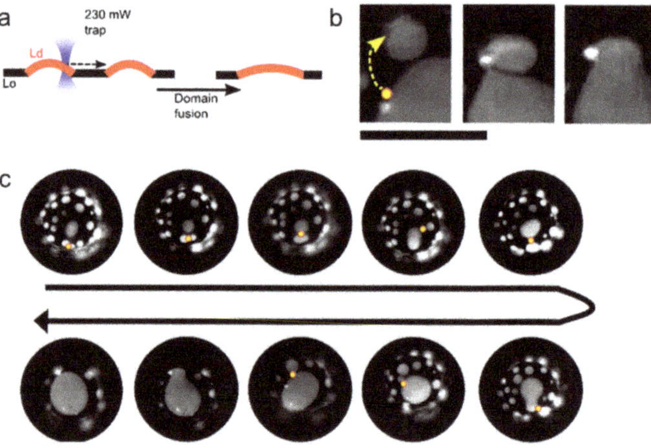

Figure 3. Controlled manipulation of membrane domains using optical traps. (**a**) Schematic showing manipulation of liquid-disordered (L_d) domains within a liquid-ordered (L_o) matrix to induce domain fusion. (**b**) Fluorescence image of a portion of a vesicle membrane showing an L_d domain being dragged by an optical trap (yellow dot and arrow) towards an adjacent domain to induce fusion. Lipid composition: DOPC/EggSM/Chol 1:1:1. (**c**) Time-course image showing eight sequential fusion events that were user-defined through domain manipulation with an optical trap (yellow circle). Scale bars = 5 μm.

2.3. Morphological Transitions

Vesicles tended to go through a series of common morphological transitions during the fusion process, which can give us insight into the underlying biophysical processes at play. The most common of these are described below, together with speculation regarding their origins.

The first relates to a morphological change during vesicle adhesion, where the vesicles 'zip-up' to form a straight interface membrane. When vesicles are generated, there is a large vesicle-to-vesicle variation in the surface tensions, depending on how much lipid material ends up in the final structure. On occasions where there is excess membrane surface area, protruding tubules are observed. When such vesicles are brought into contact with other vesicles, the tubules disappear, and they are quickly (<100 ms) subsumed into the main membrane body after adhesion (Figure 4a; Video S2). This is because, during adhesion, the total vesicle surface area increases as it deforms away from a spherical geometry. This leads to a corresponding increase in vesicle tension, as the encapsulated volume remains the same, leading to a retraction of the tubule. Assuming identical vesicle size, the ratio of the vesicle surface area increase due to membrane adhesion over the original vesicle area is given by

$$f(\vartheta) = \frac{3 - \cos\vartheta}{2^{2/3}\,(1 + \cos\vartheta)^{1/3}\,(2 - \cos\vartheta)^{2/3}} - 1 \tag{1}$$

In our experiments, the contact angle ϑ, defined as the arcsine of the adhesion patch diameter normalised with respect to the vesicle diameter, can vary between a few degrees up to c. 40°, thus resulting in relative increases of vesicle surface area of c. 1%. With a typical elastic expansion area modulus of order 200 mJ/m^2, the vesicle membrane tension can increase by up to few tens of μN/m, which can be enough to destabilise the nanotubules that protrude from the vesicle surface.

After adhesion, the vesicles were fused with one another by exposing the vesicle/vesicle interface to a continuous laser (>150 mW). Different behaviours were observed in iso-osmotic and osmotically deflated vesicles (Figure 4b). In iso-osmotic conditions (0.75 M sucrose internally, 0.35 M glucose 0.2 M NaCl externally), clean fusion events occurred 82% of the time (n = 50) with a single post-fusion vesicle formed almost instantaneously (<100 ms) after exposure to the laser (Video S3). In contrast, when the vesicles were osmotically imbalanced (0.75 M sucrose internally, 0.4 M glucose and 0.2 M NaCl externally) and, hence, had lower membrane tensions, disorderly fusion morphologies were always observed (Figure 4b, Video S4). These took the form of pearling instabilities, where tens of smaller vesicles are generated (and retained inside) the post-fusion vesicle.

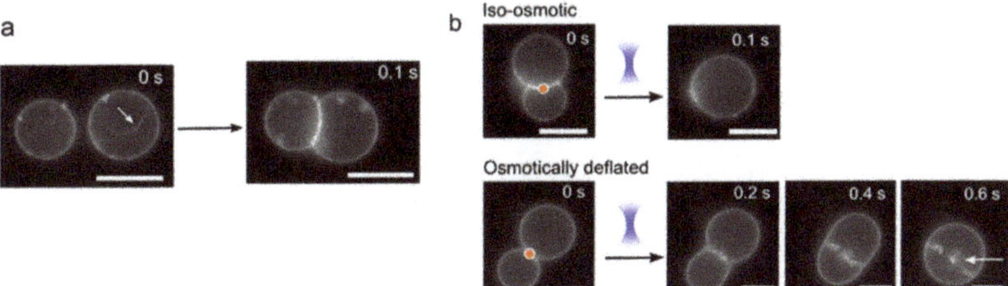

Figure 4. Vesicle morphology changes under different osmotic conditions. (**a**) If vesicles have excess membrane area and low tension, they may contain tubules (white arrow). Upon adhesion, the vesicle deform to a non-spherical geometry, which increases its tension and leads to a retraction of the tubule into the main vesicle body. (**b**) Images showing post-fusion morphologies under iso-osmotic (0.75 M sucrose internally; 0.35 M glucose, 0.2 M NaCl externally) and osmotically deflated (0.75 M sucrose internally; 0.4 M glucose, 0.2 M NaCl externally; low tension) conditions. In the former, a clean-merge is seen; in the latter the vesicle slowly relaxes to a spherical shape and many smaller internal vesicles are produced and retained inside (white arrow). Orange dot corresponds to area of laser focus. Scale bars = 5 μm.

One possible explanation for this is the creation of high energy exposed bilayers following laser exposure. Because of the low-tension and flaccid nature of the membranes, these are free to deform and reseal quickly with membrane segments that are in close proximity. This would lead to the formation of further exposed membranes in an iterative process that results in the generation of several internal vesicles. In principle, the internal membrane structures may be connected by small membrane tubules that are below the resolution of the microscope. Furthermore, the merging of two vesicles of identical size results in a post-fusion vesicle whose membrane area is c. 20% less than the combined membrane area of the two original vesicles due to the conservation of the encapsulated volume. This area reduction combined with the relatively large membrane excess area of the two deflated merging vesicles might promote the formation of membrane structures other than a single larger post-fusion vesicle, such as smaller vesicles and nanotubules.

However, several interesting 'disorderly' fusion events were also occasionally observed in iso-osmotic conditions (Figure 5). There were instances when intermediate bodies formed that were stable for c. 4 s before collapsing to generate internal membrane partitions (Figure 5a; Video S5). The intermediate structures appeared to be membranes that possessed a visible hole >3 μm in diameter. The hole resealed when the open membrane edges travelled along the surface of the vesicle, minimising the size of the pore, until the two edges met and then formed a complete internal membrane. In other instances, if the laser was continuously left on, multiple sequential opening and re-sealing events were seen. In this case, one of the vesicles would get smaller over time, until only a single unified vesicle was left (Figure 5b, Video S6). Finally, there were occasions where visible membrane pores would appear, followed by membrane rearrangement in order to yield a single internal vesicle encapsulated in a larger one (Figure 5c, Video S7).

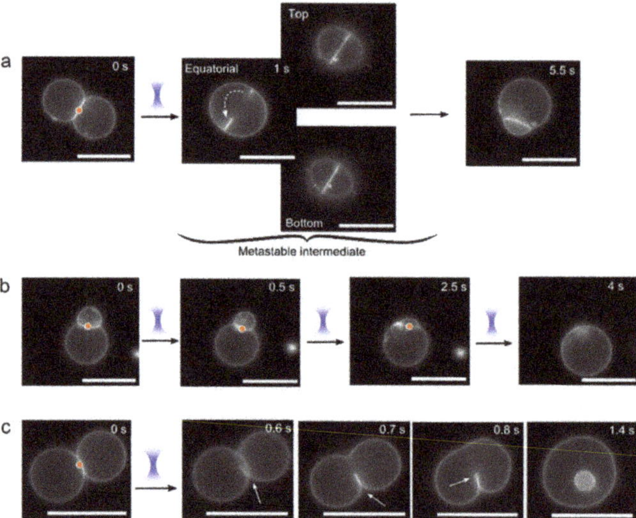

Figure 5. Post-fusion morphologies and intermediate structures. Under osmotically balanced conditions (0.75 M sucrose internally; 0.35 M glucose and 0.2 M NaCl externally) several alternative non-clean fusion events were occasionally observed after application of the laser (red dot and blue cone), including: (**a**) A meta-stable pore of c. 3 μm diameter in the bilayer partition, which after c. 5s rearranged (dotted arrow) to reform the interface membrane. (**b**) Quick opening and closing of a pore upon continuous laser illumination, after each event a portion of the membrane of one vesicle was subsumed into the other, with the formation of a single vesicle at the end. (**c**) An exposed pendant interface membrane (white arrow), which rearranged and reconnected with itself to yield in an inner vesicle. Scale bar = 5 μm.

3. Materials and Methods

3.1. Optical Trapping and Microscopy Setup

An inverted epi-fluorescence dual-carousel microscope (Nikon TE2000-U, Nikon Instruments, Tokyo, Japan) combined with an optical trapping system was used for imaging and optical manipulation, as described previously [12,28]. Briefly, the setup involves a conventional single beam optical trap that was based on a linearly polarised beam from an Ytterbium fibre laser source (20 W at 1070 nm; IPG Photonics, Europe). A bespoke 30 mm cage optic and filter cube mount was machined and replaced the upper carousel of the microscope. The laser collimator head was mounted in a cage plate and the beam was expanded to fill the back aperture of a 60 × 1.4 NA oil-immersion objective with a pair of opposing plano-convex lenses in a Keplerian arrangement along a cage optic rail. The IR dichroic mirror and IR filter (Chroma, Bellows Falls, VT, USA) were fitted in a filter cube that was mounted in the bespoke mount. The laser power at the back aperture was measured to be 9.9 ± 0.2% of the nominal laser launch power. The laser focal point was aligned to coincide with the object plane and objects were manipulated by the translation of the motorised microscope XY stage and the objective along the z-axis. The sample was imaged while using the CCD camera (ORCA-ER Hamamatsu, Shizuoka, Japan). A TRITC Nikon filter cube and a mercury-fibre illuminator (Nikon Intensilight CHGFIE, Tokyo, Japan) were used for imaging the vesicles in fluorescence mode (50 ms exposure). Image acquisition was controlled using a customised Labview (National Instruments Corp, TX, USA) interface. We note that, in all our experiments, we manipulated vesicles one at a time using a single laser. Future work concerning more advanced applications in which simultaneous trapping of multiple objects are needed will require modifications to the setup, for example, through the integration of holographic optical tweezers or time-sharing optical traps systems.

3.2. Vesicle Generation

All of the lipids were purchased from Avanti Polar Lipids (Alabaster, AL, USA) and, unless otherwise specified, reagents purchased from Sigma Aldrich (Gillingham, UK). Unless otherwise stated, all of the vesicles were composed of 1-palmitoyl-2-oleoyl-glycero-3-phosphocholine (POPC). For domain experiments, other lipids used were 1,2-dioleoyl-sn-glycero-3-phosphocholine (DOPC), Egg Sphingomyelin (EggSM), and Cholesterol (Chol). GUVs were formed via electroformation in a sucrose solution in DI water (0.75 M, unless otherwise specified in the text). Lipid solutions containing various ratios of different lipids (given in the text) were prepared together with 1 mol% fluorescent lipid 1,2-dioleoyl-sn-glycero-3-phosphoethanolamine-N-(lissamine rhodamine B sulfonyl) (RhPE). These were prepared by dissolving appropriate amounts of lipid in chloroform to yield a 1 mg mL^{-1} solution. 40 µL of this solution was then spread evenly on an indium tin oxide (ITO) slide, and a lipid film was deposited as the chloroform evaporated. The slide was placed in a desiccator for a minimum of 30 min to remove residual chloroform. A 5 mm thick polydimethylsiloxane (PDMS) spacer with a central cut-out was sandwiched in between two ITO slides, one of which contained the lipid film, with the conductive sides facing each other. This chamber was held together with clips, and it was filled with the 0.75 M sucrose solution. An alternating electric field (1 V, 10 Hz) was applied across the ITO plates while using a function generator (Aim-TTi, TG315, Huntingdon, UK). The sample was left to electroform in a 60° oven to ensure that the lipid was in the fluid phase. After two hours, the electric field was changed to 1 V, 2 Hz for a further hour, and the resulting vesicles collected.

3.3. Fusion Experiments

For vesicles fusion, there needed to be AuNPs present at the vesicle-vesicle interface. This was achieved by functionalizing the vesicles by adding 2 wt.% 16:0 Biotinyl Cap PE to the initial lipid film. After vesicle formation, 150 nm streptavidin-coated AuNPs (Nanopartz, CO, USA; product C11-150-TS-50; 2.5 mg·mL^{-1}) were added to the vesicles 1:9, with the sample being vortexed for 30 min to drive conjugation. For the fusion experiments, the vesicles were deposited on a coverslip coated with a BSA monolayer to prevent interactions between the glass substrate and the vesicle membrane. The coating was applied by depositing 300 μL of 1% BSA in DI water on a coverslip and leaving it to evaporate in a 60 °C oven, leaving behind a protein film. The film was subsequently rinsed with DI water and dried under a nitrogen stream. The vesicle assembly chambers were prepared by placing a 1 mm thick PDMS sheet with a 10 mm diameter hole on the coverslip.

For the fusion experiments, unless otherwise stated, all of the experiments were conducted under iso-osmotic conditions. A sample of vesicles with 0.75 M sucrose internally, and 0.35 M glucose, 0.2 M NaCl externally were created by diluting the vesicles following electroformation with appropriate volumes of NaCl and glucose solutions, all in DI water. The sample was then mixed by pipette aspiration, the chamber sealed with a second coverslip, and then placed on the optical trapping setup. Individual vesicles were trapped by switching the laser on (20 mW–100 mW at trap), and were moved in three-dimensional (3-D) relative to the sample by moving the microscope stage (x,y) and changing the focus of the objective (z). These relatively low powers were needed to avoid unintended fusion. Once the vesicles were brought into contact with one another to form an adhesion patch, fusion was initialised by applying a laser power of >150 mW (at trap) at the membrane patch.

4. Conclusions

We have conducted a series of feasibility experiments to demonstrate the potential of laser-assisted membrane manipulation. Firstly, we show that optical trapping technologies can be used to dynamically move across a membrane phase diagram by changing the composition of a bilayer through the delivery of new membrane material. This is powerful, as it paves the way for systematic changes to the lipid composition of the membrane while still retaining the encapsulated cargo. It also enables the study of non-equilibrium dynamic processes, such as the kinetics of membrane rearrangement after perturbations in lipid compositions are imposed. Unlike alternative methods (e.g., using reagents that remove specific membrane molecular constituents) [29], this is a generic method, which, in principle, allows for the introduction of any material (lipid, cholesterol, protein, etc.) embedded in vesicles, as well as the controlled addition of non-biological material (e.g., block copolymers, dendrimers etc.) in controlled quantities. Similar approaches based on electrofusion have been shown to bypass issues that were associated with batch variation of vesicle lipid composition by changing the lipid composition post-vesicle generation [18]. Despite the laser-induced fusion, NP-labelled membranes may result in relatively larger temperature (>100 °C) that could compromise the functionality of biological material; such a temperature increase is localised in a small region of a size that is comparable to the NP diameter (100 nm). Since outside this region the temperature decreases as the inverse of the distance, it is expected that the vast majority of the cargo remains unaffected.

Secondly, we show the controlled fusion of coexisting domains in a vesicle. This can unlock applications for controlled chemical reactions on a membrane surface, for example, if reactive species can be bound to defined domains, and it can be deployed in future in studies that concern the biophysics of domain coalescence.

Finally, the fine spatiotemporal control that is associated with the use of laser has meant that we were able to observe a repertoire of membrane transformations, including: (i) the formation of large open membrane intermediate structures that appear to be metastable; (ii) the opening/resealing of transient pores (<1 s); and, (iii) the generation of internal daughter vesicles post-fusion. Many of these could be repurposed in a synthetic biology context. When coupled with microfluidic [30] and related technologies [31,32], it could be used to sculpt new cell-mimetic motifs (e.g., synthetic nuclei), replicate cellular processes (e.g., endocytosis), and gain insight into cell membrane biophysics in a model environment. Taken together, these results serve as a feasibility roadmap to guide future detailed studies and exploitation of laser-assisted membrane manipulation technologies.

Supplementary Materials: The following are available online at http://www.mdpi.com/2072-666X/11/4/388/s1: Video S1: Controlled optical manipulation and fusion of membrane domains. Video S2: Post-adhesion vesicle deformation and retraction of the tubule. Video S3: Clean vesicle fusion under iso-osmotic conditions. Video S4: Disorderly vesicle fusion under osmotically deflated conditions. Video S5: Transient visible membrane pores post-fusion. Video S6: Multiple pore opening and closing events upon laser exposure. Video S7: Post-fusion generation of a single internal daughter vesicle.

Author Contributions: Y.E. devised the experimental concepts and led the project. A.V., G.B., and Y.E. performed the experiments, analysis, and contributed to writing the manuscript. All authors have read and agreed to the published version of the manuscript.

Funding: This research was funded by EPSRC grant EP/N016998/1 and a UKRI Future Leaders Fellowship MR/S031537/1 awarded to Y.E.

Conflicts of Interest: The authors declare no conflict of interest. The funders had no role in the design of the study; in the collection, analyses, or interpretation of data; in the writing of the manuscript, or in the decision to publish the results.

References

1. Buddingh, B.C.; van Hest, J.C. Artificial cells: Synthetic compartments with life-like functionality and adaptivity. *Acc. Chem. Res.* **2017**, *50*, 769–777. [CrossRef] [PubMed]
2. Göpfrich, K.; Platzman, I.; Spatz, J.P. Mastering complexity: Towards bottom-up construction of multifunctional eukaryotic synthetic cells. *Trends Biotechnol.* **2018**, *36*, 938–951. [CrossRef] [PubMed]
3. Trantidou, T.; Friddin, M.; Salehi-Reyhani, A.; Ces, O.; Elani, Y. Droplet microfluidics for the construction of compartmentalised model membranes. *Lab Chip* **2018**, *18*, 2488–2509. [CrossRef] [PubMed]
4. Martens, S.; McMahon, H.T. Mechanisms of membrane fusion: Disparate players and common principles. *Nat. Rev. Mol. Cell Biol.* **2008**, *9*, 543. [CrossRef]
5. Daraee, H.; Etemadi, A.; Kouhi, M.; Alimirzalu, S.; Akbarzadeh, A. Application of liposomes in medicine and drug delivery. *Artif. Cells Nanomed. Biotechnol.* **2016**, *44*, 381–391. [CrossRef]
6. Robinson, T.; Verboket, P.E.; Eyer, K.; Dittrich, P.S. Controllable electrofusion of lipid vesicles: Initiation and analysis of reactions within biomimetic containers. *Lab Chip* **2014**, *14*, 2852–2859. [CrossRef]
7. Caschera, F.; Sunami, T.; Matsuura, T.; Suzuki, H.; Hanczyc, M.M.; Yomo, T. Programmed vesicle fusion triggers gene expression. *Langmuir* **2011**, *27*, 13082–13090. [CrossRef]
8. Csiszár, A.; Hersch, N.; Dieluweit, S.; Biehl, R.; Merkel, R.; Hoffmann, B. Novel fusogenic liposomes for fluorescent cell labeling and membrane modification. *Bioconjug. Chem.* **2010**, *21*, 537–543. [CrossRef]
9. Guo, Y.; Wu, M.; Chen, H.; Wang, X.; Liu, G.; Li, G.; Ma, J.; Sy, M.-S. Effective tumor vaccine generated by fusion of hepatoma cells with activated B cells. *Science* **1994**, *263*, 518–520. [CrossRef]
10. Peng, K.-W.; TenEyck, C.J.; Galanis, E.; Kalli, K.R.; Hartmann, L.C.; Russell, S.J. Intraperitoneal therapy of ovarian cancer using an engineered measles virus. *Cancer Res.* **2002**, *62*, 4656–4662.
11. Galanis, E.; Bateman, A.; Johnson, K.; Diaz, R.M.; James, C.D.; Vile, R.; Russell, S.J. Use of viral fusogenic membrane glycoproteins as novel therapeutic transgenes in gliomas. *Hum. Gene Ther.* **2001**, *12*, 811–821. [CrossRef] [PubMed]
12. Bolognesi, G.; Friddin, M.S.; Salehi-Reyhani, A.; Barlow, N.E.; Brooks, N.J.; Ces, O.; Elani, Y. Sculpting and fusing biomimetic vesicle networks using optical tweezers. *Nat. Commun.* **2018**, *9*, 1882. [CrossRef] [PubMed]
13. Deshpande, S.; Wunnava, S.; Hueting, D.; Dekker, C. Membrane Tension–Mediated Growth of Liposomes. *Small* **2019**, *15*, 1902898. [CrossRef] [PubMed]
14. Pantazatos, D.; MacDonald, R. Directly observed membrane fusion between oppositely charged phospholipid bilayers. *J. Membr. Biol.* **1999**, *170*, 27–38. [CrossRef] [PubMed]
15. Stengel, G.; Zahn, R.; Höök, F. DNA-induced programmable fusion of phospholipid vesicles. *J. Am. Chem. Soc.* **2007**, *129*, 9584–9585. [CrossRef] [PubMed]
16. Fan, Z.-A.; Tsang, K.-Y.; Chen, S.-H.; Chen, Y.-F. Revisit the Correlation between the Elastic Mechanics and Fusion of Lipid Membranes. *Sci. Rep.* **2016**, *6*, 31470. [CrossRef]
17. Haluska, C.K.; Riske, K.A.; Marchi-Artzner, V.; Lehn, J.-M.; Lipowsky, R.; Dimova, R. Time scales of membrane fusion revealed by direct imaging of vesicle fusion with high temporal resolution. *Proc. Natl. Acad. Sci. USA* **2006**, *103*, 15841–15846. [CrossRef]

18. Bezlyepkina, N.; Gracià, R.; Shchelokovskyy, P.; Lipowsky, R.; Dimova, R. Phase diagram and tie-line determination for the ternary mixture DOPC/eSM/cholesterol. *Biophys. J.* **2013**, *104*, 1456–1464. [CrossRef]

19. Rørvig-Lund, A.; Bahadori, A.; Semsey, S.; Bendix, P.M.; Oddershede, L.B. Vesicle fusion triggered by optically heated gold nanoparticles. *Nano Lett.* **2015**, *15*, 4183–4188. [CrossRef]

20. Bahadori, A.; Oddershede, L.B.; Bendix, P.M. Hot-nanoparticle-mediated fusion of selected cells. *Nano Res.* **2017**, *10*, 2034–2045. [CrossRef]

21. Bendix, P.M.; Reihani, S.N.S.; Oddershede, L.B. Direct measurements of heating by electromagnetically trapped gold nanoparticles on supported lipid bilayers. *ACS Nano* **2010**, *4*, 2256–2262. [CrossRef] [PubMed]

22. Friddin, M.S.; Bolognesi, G.; Salehi-Reyhani, A.; Ces, O.; Elani, Y. Direct manipulation of liquid ordered lipid membrane domains using optical traps. *Commun. Chem.* **2019**, *2*, 6. [CrossRef]

23. Biner, O.; Schick, T.; Müller, Y.; von Ballmoos, C. Delivery of membrane proteins into small and giant unilamellar vesicles by charge-mediated fusion. *FEBS Lett.* **2016**, *590*, 2051–2062. [CrossRef] [PubMed]

24. Veatch, S.L.; Keller, S.L. Separation of liquid phases in giant vesicles of ternary mixtures of phospholipids and cholesterol. *Biophys. J.* **2003**, *85*, 3074–3083. [CrossRef]

25. Karamdad, K.; Hindley, J.W.; Bolognesi, G.; Friddin, M.S.; Law, R.V.; Brooks, N.J.; Ces, O.; Elani, Y. Engineering thermoresponsive phase separated vesicles formed via emulsion phase transfer as a content-release platform. *Chem. Sci.* **2018**, *9*, 4851–4858. [CrossRef]

26. Bacia, K.; Schwille, P.; Kurzchalia, T. Sterol structure determines the separation of phases and the curvature of the liquid-ordered phase in model membranes. *Proc. Natl. Acad. Sci. USA* **2005**, *102*, 3272–3277. [CrossRef]

27. Rim, J.; Ursell, T.; Phillips, R.; Klug, W. Morphological phase diagram for lipid membrane domains with entropic tension. *Phys. Rev. Lett.* **2011**, *106*, 057801. [CrossRef]

28. Friddin, M.S.; Bolognesi, G.; Elani, Y.; Brooks, N.J.; Law, R.V.; Seddon, J.M.; Neil, M.A.; Ces, O. Optically assembled droplet interface bilayer (OptiDIB) networks from cell-sized microdroplets. *Soft Matter* **2016**, *12*, 7731–7734. [CrossRef]

29. Zidovetzki, R.; Levitan, I. Use of cyclodextrins to manipulate plasma membrane cholesterol content: Evidence, misconceptions and control strategies. *Biochim. Biophys. Acta (BBA)-Biomembr.* **2007**, *1768*, 1311–1324. [CrossRef]

30. Van Swaay, D. Microfluidic methods for forming liposomes. *Lab Chip* **2013**, *13*, 752–767. [CrossRef]

31. Friddin, M.S.; Elani, Y.; Trantidou, T.; Ces, O. New directions for artificial cells using prototyped biosystems. *Anal. Chem.* **2019**, *91*, 4921–4928. [CrossRef] [PubMed]

32. Carreras, P.; Elani, Y.; Law, R.; Brooks, N.; Seddon, J.; Ces, O. A microfluidic platform for size-dependent generation of droplet interface bilayer networks on rails. *Biomicrofluidics* **2015**, *9*, 064121. [CrossRef] [PubMed]

 *micromachines*

Article

Single-Molecule Mechanics in Ligand Concentration Gradient

Balázs Kretzer, Bálint Kiss, Hedvig Tordai, Gabriella Csík, Levente Herényi and Miklós Kellermayer *

Department of Biophysics and Radiation Biology, Semmelweis University, 1094 Budapest, Hungary; kretzer.b1@gmail.com (B.K.); onlybalint@gmail.com (B.K.); tordaih@hegelab.org (H.T.); csik.gabriella@med.semmelweis-univ.hu (G.C.); herenyi.levente@med.semmelweis-univ.hu (L.H.)
* Correspondence: kellermayer.miklos@med.semmelweis-univ.hu; Tel.: +36-20-825-9994

Received: 28 December 2019; Accepted: 14 February 2020; Published: 19 February 2020

Abstract: Single-molecule experiments provide unique insights into the mechanisms of biomolecular phenomena. However, because varying the concentration of a solute usually requires the exchange of the entire solution around the molecule, ligand-concentration-dependent measurements on the same molecule pose a challenge. In the present work we exploited the fact that a diffusion-dependent concentration gradient arises in a laminar-flow microfluidic device, which may be utilized for controlling the concentration of the ligand that the mechanically manipulated single molecule is exposed to. We tested this experimental approach by exposing a λ-phage dsDNA molecule, held with a double-trap optical tweezers instrument, to diffusionally-controlled concentrations of SYTOX Orange (SxO) and tetrakis(4-N-methyl)pyridyl-porphyrin (TMPYP). We demonstrate that the experimental design allows access to transient-kinetic, equilibrium and ligand-concentration-dependent mechanical experiments on the very same single molecule.

Keywords: optical tweezers; concentration gradient; force spectroscopy; diffusion; microfluidics; fluorescence

1. Introduction

Single-molecule methods, which have been evolving progressively in the past thirty years [1–11], give unprecedented insights into mechanistic details of molecular phenomena because they provide the distribution of parameters beyond ensemble averages, reveal stochastic processes such as fluorescence blinking [12], uncover trajectories of processes that evolve along parallel pathways such as protein folding [13–15], and allow the characterization of mechanical functions and properties such as molecular elasticity and motor-enzyme force generation [16–18]. One of the major tools in single-molecule mechanics is optical tweezers [1,19–21], which have been successfully employed in the investigation of DNA [22,23], RNA [24] and proteins [13,15]. An important challenge in single-molecule investigations is to provide a suitable, possibly controllable, chemical environment. Optical tweezers may be combined with microfluidics to provide discrete changes in the solution conditions [25,26]. However, exposing a single molecule to continuously controlled concentration of solutes remains a challenge.

In the present work, we utilize the concentration gradient that evolves in a parallel, laminar-flow multichannel microfluidic device, to expose a model DNA molecule to controlled concentrations of model intercalators SYTOX Orange (SxO) and a porphyrin derivative (TMPYP). Such DNA-binding small molecules have long been of interest due to their widespread applications. They are often used as therapeutic drugs [27–32] or as building blocks in creating functional assemblies [33–37]. They are most frequently used in molecular biology as DNA-labeling agents that can also perturb enzymatic reactions [38]. They bind to DNA in several modes: major- and minor-groove binding, electrostatic/allosteric binding and intercalation [39]. A molecule may bind with multiple modes depending on the nucleotide sequence [40,41], and each mode may have different equilibration times [2]. The characteristic time for association and dissociation may vary extensively, ranging from less than a second to hours [38]. Thus, for a precise characterization of the chemical reactions under mechanical loads it is necessary to control the experimental parameters of force, ligand concentration and exposure time simultaneously and precisely. We show that by positioning a pre-stretched single λ-phage dsDNA molecule at different locations within a diffusionally controlled spatial concentration gradient of the DNA-binding ligand, the parameters of the binding reaction can be accessed by following the concentration-dependent changes in DNA extension. Furthermore, force-driven structural changes in the very same DNA molecule can be measured at well-controlled ligand concentrations.

2. Materials and Methods

2.1. Sample Preparation

Biotinylated oligonucleotides (Sigma-Aldrich, St. Louis, MO, USA) were used to label the 3′ recessive ends of λ-phage dsDNA (Thermo Fisher Scientific, Waltham, MA, USA). Ligation was carried out with T7 DNA ligase (Thermo Fisher Scientific). Following biotin labeling the DNA samples were stored at 4 °C to prevent degradation caused by freezing [42]. For optical tweezer measurements DNA was used at 30 ng/mL concentration in Tris-NaCl buffer (20 mmol/L Tris-HCl, pH 7.4, 50 mmol/L NaCl) throughout the experiments. DNA molecules were tethered between 3.11 μm diameter streptavidin-coated polystyrene beads (Kisker Biotech, Steinfurt, Germany). As model ligands that bind DNA we used the fluorescent mono-intercalating dye SYTOX Orange (SxO) (Thermo Fisher Scientific) and tetrakis(4-N-methyl)pyridyl-porphyrin (TMPYP).

2.2. Single-Molecule Manipulation and Imaging

Single molecules of λ-phage DNA were mechanically manipulated by using a C-Trap dual trap optical tweezers system equipped with a multi-channel microfluidic flow-through sample chamber system and laser-scanning confocal fluorescence microscopy (Lumicks, Amsterdam, The Netherlands). The captured molecule was positioned at a distance ~100 μm away from the microchannel confluence to ensure a stable hydrodynamic environment. A typical solution flow rate of 1 mm/s was maintained throughout the experiments. Mechanical measurements were carried out in either constant stretch velocity (500 nm/s) or constant force (35 pN) modes. To measure fluorescence, a 532 nm YAG laser (Lumicks) was used for excitation. Fluorescence emission, across a wavelength range of 545–620 nm, was recorded with a photon-counting avalanche photodiode. We collected images either in the X–Y space, or kymograms in the molecular distance–time space.

2.3. Calibration of Concentration Gradient with Fluorescence

A diffusion-controlled ligand concentration gradient was generated between two, neighboring microfluidic channels, one of which contained high concentration of the DNA-binding molecule, and the other one containing buffer only. The shape of the concentration gradient depended on the limiting ligand concentrations ($c_0 = 0$ and $c_s = 100$ nmol/L), and the diffusion time (t) set by the flow velocity ($v = 1000$ μm/s) and the distance from the microchannel confluence ($d \sim 100$ μm) as $t = d/v$ (see also Section 2.5 below) [43]. The actual spatial distribution of ligand concentration was characterized by imaging SxO fluorescence intensity along the gradient (i.e., in perpendicular to the flow direction) (Figure 1). The intensity profile (fluorescence intensity vs. distance) was calculated from the acquired image with the ImageJ software (version 1.52, public domain).

2.4. Molecular Scanning of the Ligand Concentration Gradient

Considering that the investigated ligands (SxO and TMPYP) intercalate into DNA and hence affect DNA's contour length [38,40], the ligand concentration gradient may be mechanically mapped by moving a captured DNA along the gradient direction. We used two different methods: in the first, a single λ-phage dsDNA molecule, pulled taut with a constant force of 35 pN, was moved with constant velocity (18.5 µm/s) along the gradient (Figure 1). The molecular axis was parallel with the flow direction. Normalized length change was plotted against the distance traveled. In the second method, the dsDNA molecule was rapidly moved (with a speed of 500 µm/s) in discrete, 38 µm steps along the gradient. We plotted dsDNA length as a function of time. Whereas the first approach mimics an equilibrium experiment, the second approach is a transient kinetic one.

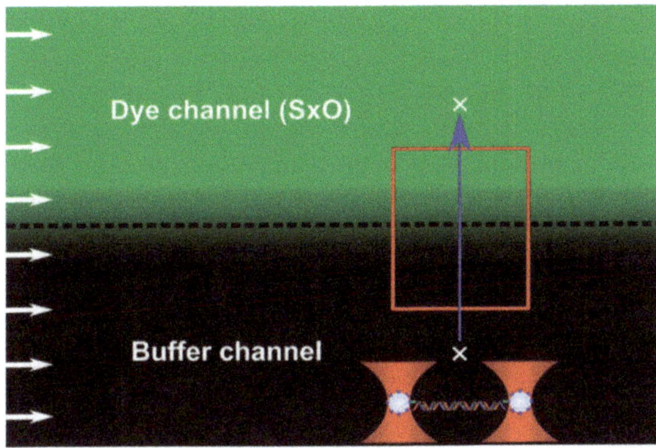

Figure 1. Schematics of the experimental design in the microfluidic system. White arrows on the left indicate constant flow velocity in the microchannels (dye and buffer channels). The dashed line marks the theoretical border between the two channels. Sytox Orange (SxO) concentration gradient arises due to diffusion between the neighboring microchannels, with the green background color corresponding to SxO fluorescence intensity hence the concentration. For the sake of simplicity, the progressive decay in the concentration gradient along the flow direction is neglected. The red rectangle indicates the area sampled either by the fluorescence or mechanical measurements. The blue arrow marks the path of the tethered DNA molecule, held by its ends with beads captured in independent optical traps (bottom of image), moved along the concentration gradient. White crosses mark the start and end positions of the mechanical trajectory.

2.5. Theory

The evolution of the concentration gradient at the border between the neighboring microchannels is dictated by diffusion according to Fick's second law, a one-dimensional solution of which is:

$$c(x,t) = \frac{c_s}{2}\left[1 + erf\left(\frac{x}{2\sqrt{Dt}}\right)\right] = \frac{c_s}{2}\left[1 + erf\left(\frac{x}{\sqrt{2}\bar{x}}\right)\right],$$

(1)

where c is diffusant concentration, c_s is the maximum, initial concentration in the dye channel, D is the diffusion coefficient, t is time, and x is distance along the one-dimensional diffusion coordinate (in perpendicular to the flow direction) so that x is 0 at the microchannel border and positive towards the dye channel. The parameter *erf* is an error function in the form of an integral of a Gaussian [44]. Considering that according to Einstein's theory of diffusion the root-mean-square displacement of Brownian particle is $\bar{x} = \sqrt{2Dt}$, the term $2\sqrt{Dt}$ in Equation (1) may be replaced with $\sqrt{2}\bar{x}$ [45]. By removing D from the equation, we neglect the role of the diffusion coefficient, but we note that differences in the diffusibility of the different ligands alter the final shape of their respective concentration gradients. The boundary conditions of Equation (1) are such that the concentration gradients at infinite distances away from the microchannel border in either direction are 0. The initial conditions of Equation (1) are such that $c(x,0)$ is 0 at $x < 0$ and c_s at $x \geq 0$ [44]. We used Equation (1) to fit the fluorescence intensity vs. distance profile across the microchannel border.

3. Results and Discussion

In this work we investigated the applicability of a diffusionally generated concentration gradient in single-molecule biophysics. The concept of the experimental layout is shown in Figure 1. In the employed instrument force-measuring double-trap optical tweezers were combined with laser scanning confocal fluorescence imaging and laminar-flow microfluidics. Whereas the optical tweezers and fluorescence imaging allowed mechanical manipulation (molecule positioning, stretching, relaxing) and imaging, respectively, the multichannel microfluidic device enabled efficient solution control and the generation of a ligand concentration gradient at the border of vicinal microchannels. In the present experiments we manipulated λ-phage dsDNA as a model molecule, and used SYTOX Orange [38] and a porphyrin derivative (TMPYP) [34,41] as model DNA-binding ligands. To develop the ligand concentration gradient, one of the two neighboring laminar-flow microchannels ("dye channel" in Figure 1) contained high concentration of the ligand, whereas its neighbor only buffer ("buffer channel"). Such concentration gradients have in the past been employed for controlling cell behavior [46–48]. We note that in order to develop a concentration gradient in which the manipulated molecule responds across the entire concentration scale, the ligand concentration in the dye channel should be below the saturating concentration. Here, we used ligand concentrations of 100 nmol/L in the dye channel, which are close to saturation levels [38,40,41]. The concentration gradient develops as a function of time as the neighboring solutions flow, in parallel, with pre-adjusted velocity, down the microfluidic device. The shape (i.e., steepness) of the gradient can thus be chosen either by adjusting the flow velocity and/or by selecting the sampling position downstream of the microchannel confluence. We used a flow velocity of 1 mm/s throughout our experiments, and the sample (i.e., the trapped DNA molecule) was positioned at about 100 µm distal of the confluence.

We characterized the ligand concentration gradient by imaging SxO fluorescence and by mechanically scanning the gradient with the captured DNA molecule (Figure 2). Fluorescence imaging of the sampled microchannel border region (Figure 2A) revealed an area in which fluorescence intensity gradually increased from the buffer channel towards the dye channel. The corresponding intensity profile plot (Figure 2B) displayed a sigmoidal function which could be well fitted with Equation (1). The fit yielded 23.3 ± 0.1 µm for diffusion length, indicating that at the employed settings (i.e., maximum ligand concentration, flow velocity, distance from confluence) approximately 46 µm distance is available across which ligand concentration varies essentially linearly between 20–80 nmol/L. Mechanical scanning of the concentration gradient gave slightly different results (Figure 2C). It is possible to mechanically scan the ligand concentration gradients with dsDNA, because the binding of intercalating molecules results in DNA lengthening [38]. Accordingly, the DNA lengthening (normalized length increment) vs. distance (along concentration gradient) function displayed sigmoid curves for both SxO and TMPYP, although their trajectories were slightly different from that of fluorescence. In these experiments the DNA molecule was first pulled taut with a force of 35 pN (kept constant using feedback) so that its end-to-end length approximated its contour length, and the scanning of the gradient was carried out with a slow speed of 18.5 µm/s to approach chemical equilibrium. Fitting the SxO data with Equation (1) yielded 29.0 ± 0.2 µm for diffusion length, indicating that based on mechanical scanning about 58 µm distance is available across which ligand concentration varies essentially linearly between 20–80 nmol/L. Thus, mechanical scanning provides a more sensitive method of mapping the concentration gradient than fluorescence imaging with the current instrumentation settings. For practical purposes, an initial mechanical scanning of the concentration gradient with a DNA molecule is recommended, so that subsequent single-molecule mechanics experiments can be carried out at the desired ligand concentration. Altogether, these experiments provide the equilibrium length of the DNA molecule at the specific ligand concentration and at the specific, pre-adjusted force. The different shape of the extension-concentration trace observed in the case of TMPYP points at different binding mechanisms of TMPYP to DNA, which warrants further, detailed experimental exploration.

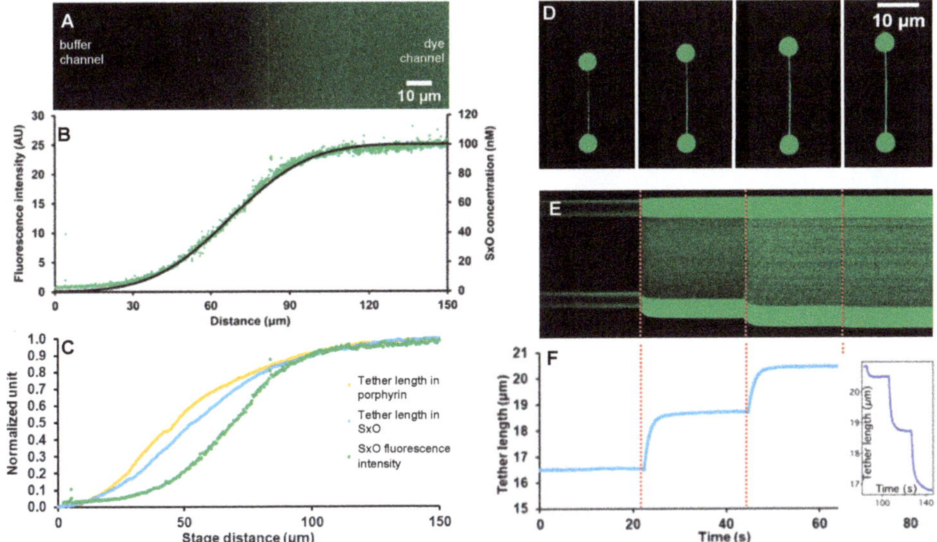

Figure 2. Characteristics of the ligand concentration gradient. (**A**) Laser scanning confocal microscopic image of the sampled microfluidic device area, across neighboring microchannels. The buffer and dye (SxO) channels are towards the left and right of the image, respectively. Initial SxO concentration (towards the right) is 100 nmol/L. Green coloring is artificial. (**B**) Fluorescence intensity (in arbitrary units, A.U.) measured along the SxO concentration gradient (green dots). Equation (1) was used to fit the experimental data (black continuous line). (**C**) Normalized DNA lengthening caused by mechanically sampling SxO (blue) and TMPYP (tetrakis(4-N-methyl)pyridyl-porphyrin, yellow) gradients by moving the DNA molecule, pulled taut with a constant 35 pN force, with a constant speed of 18.5 μm/s. For reference, the SxO fluorescence intensity data are also shown (green). (**D**) Images of DNA molecules, held stretched between two microbeads with constant force (35 pN), at four different positions (19, 57, 95, 133 μm, from left to right) along the SxO concentration gradient. The upper bead was re-positioned by the feedback system in order to maintain constant force. (**E**) Kymogram obtained by confocal scanning along the axis of the stretched DNA as a function of time during stepwise (38 μm/step) translation of the molecule along the SxO concentration gradient. Red dashed lines indicate the time points when the 38 μm steps were made. (**F**) DNA tether length as a function of time during rapid (500 μm/s), stepwise translation of the molecule along the SxO concentration gradient. In control experiments lacking SxO the thether length stayed constant. Inset shows the relaxation of tether length during stepwise translation of the same DNA molecule through the same spatial positions in the SxO concentration gradient. The non-linear decay curves, collected either towards or backwards the gradient, were fitted with the single-exponential function $L = L_0 + \Delta L e^{-t/\tau_{eq}}$, where L and L_0 are the actual and starting tether lengths, respectively, ΔL is the maximal tether-length change, t is time and τ_{eq} is the equilibration time constant.

The use of SxO, a fluorescent intercalating dye, allows the direct imaging of the dsDNA molecule as it is translated across the concentration gradient (Figure 2D). The stark contrast between the stained DNA molecule and the microfluidic background emerges due to a ~450 fold increase of SxO fluorescence emission upon DNA binding [49]. As the captured DNA molecule was rapidly (with a speed of 500 μm/s) advanced to a new location along the SxO concentration gradient, its average fluorescence intensity increased and its contour length relaxed to an increased level (Figure 2E). Considering that at the employed speed the travel along the 38 μm step distance took only 76 ms, the subsequent lengthening, which relaxed fully in about five seconds (Figure 2F), corresponds to the kinetics of chemical equilibration rather than to the mechanical perturbation caused by the translational motion.

Fitting single-exponential functions to the consecutive lengthening phases (Figure 2F) thus gave chemical equilibration time constants (τ_{eq}) of 1.09, 1.22 and 1.23 s. In the reverse process, during which the DNA molecule was translated down the SxO gradient to the same spatial positions, we observed contraction relaxation steps (Figure 2F, inset). Fitting single-exponential functions to the consecutive contraction phases gave τ_{eq} of 1.18, 1.18 and 1.68 s. Considering that in the last step the DNA molecule was positioned in an environment devoid of SxO, the equilibration time constant of this reaction corresponds to τ_{off}. Notably, the obtained value (1.68 s) compares well to that measured earlier for the interaction of SxO with DNA at this force level [38]. Altogether, these experiments demonstrated that by combining optical tweezers with a pre-adjusted ligand concentration gradient a range of experiments can be carried out on the same single molecule, thereby providing rapid access to numerous equilibrium and kinetic parameters of a chemical reaction.

Our experimental setting allows the exploration of ligand concentration-dependent effects on the mechanical properties of a filamentous molecule. We investigated how increasing concentrations of SxO affect the force-extension curve of dsDNA by stretching and relaxing the same λ-phage DNA molecule at different spatial locations in the SxO gradient (Figure 3). Upon increasing SxO concentration the DNA molecule became longer (the non-linear force trace in the entropic regime shifted to greater lengths), and the cooperative overstretch transition and force hysteresis gradually disappeared in the explored extension range. The results show that the majority of the complex set of changes occurs below an SxO concentration of 15 nmol/L. By varying the parameters for generating the concentration gradient (flow velocity, ligand concentration in the dye channel, distance of sample location from microchannel confluence), mechanical measurements can be carried out across specific, narrow ligand concentration regimes, thereby uncovering further details of the molecular mechanisms of intercalator-DNA interactions.

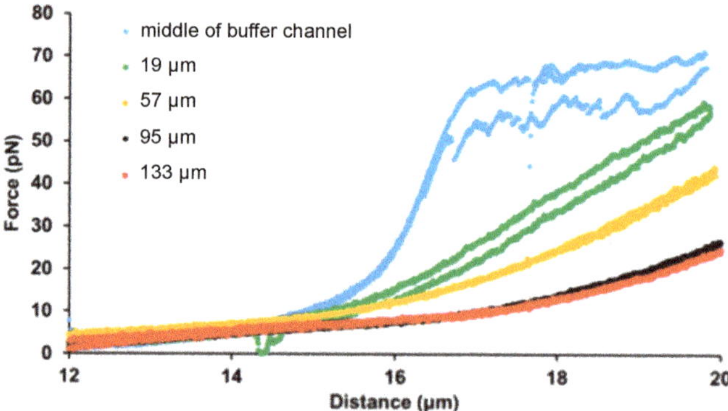

Figure 3. Constant-velocity stretch–relaxation cycles of DNA in a SxO concentration gradient. Force vs. extension curves for the same λ-phage dsDNA molecule were measured at five different locations of the microfluidic device (see legend): middle of the buffer channel (blue), and 19 (green), 57 (yellow), 95 (black), 133 (red) µm along the concentration gradient. The positions correspond to approximate SxO concentrations of 0, 15, 70, 95, and 98 nmol/L.

4. Conclusions

In conclusion, in the present work we successfully employed diffusionally generated concentration gradients of DNA-binding ligands for measuring concentration-dependent effects on single-molecule mechanics. The combination of the double-trap optical tweezers with fluorescence imaging and carefully designed microfluidic environment allowed for rapid and efficient measurement of the mechanical properties of DNA exposed to different ligand concentrations of intercalators and provided access to numerous equilibrium and kinetic parameters of the ligand-DNA interaction. Our initial measurements on the comparison of SxO and TMPYP suggested that the binding of TMPYP to DNA is likely to follow mechanisms that are different from or additional to those of SxO. The major advantage of the use of the concentration gradient is that uncertainties of ligand concentration adjustment caused by the adsorption and desorption of cyanine dyes to and from the surfaces of the microfluidic device [38] can be alleviated. The concentration gradient may also be employed for investigating the effects of ligands (e.g., ions, substrates, etc.) on proteins. Although the small size of usual proteins may pose experimental challenge, large filamentous proteins (e.g., titin [50]) or proteins captured with DNA handles [51–54] may be investigated directly. Considering the emerging significance of single-molecule mechanics in understanding structural and functional detail, the addition of precisely adjusted ligand concentration gradients, as demonstrated here, may provide further access to understanding the exact mechanisms behind biomolecular phenomena.

Author Contributions: B.K. (Balázs Kretzer), B.K. (Bálint Kiss) and M.K. conceived the experiments, B.K. (Balázs Kretzer) and B.K. (Bálint Kiss) performed measurements, H.T. and G.C. provided reagents, B.K. (Balázs Kretzer), B.K. (Bálint Kiss), L.H. and M.K. analyzed data, B.K. (Balázs Kretzer), B.K. (Bálint Kiss), H.T., G.C., L.H. and M.K. wrote and edited the manuscript. All authors have read and agreed to the published version of the manuscript.

Funding: This research was funded by grants from the Hungarian National Research, Development and Innovation Office (K124966; National Heart Program NVKP-16-1-2016-0017; Thematic Excellence Programme; National Bionics Programme ED_17-1-2017-0009).

Acknowledgments: We thank Mónika Komárné Drabbant and Krisztina Szendefyné Lór for providing technical assistance.

References

1. Block, S.M. Making light work with optical tweezers. *Nature* **1992**, *360*, 493–495. [CrossRef] [PubMed]
2. Bustamante, C.; Bryant, Z.; Smith, S.B. Ten years of tension: Single-molecule DNA mechanics. *Nature* **2003**, *421*, 423–427. [CrossRef] [PubMed]
3. Su, Q.P.; Ju, L.A. Biophysical nanotools for single-molecule dynamics. *Biophys. Rev.* **2018**, *10*, 1349–1357. [CrossRef] [PubMed]
4. Bustamante, C.; Macosko, J.C.; Wuite, G.J.L. Grabbing the cat by the tail: Manipulating molecules one by one. *Nat. Rev. Mol. Cell Biol.* **2000**, *1*, 130–136. [CrossRef] [PubMed]
5. Bustamante, C.; Smith, S.B.; Liphardt, J.; Smith, D. Single-molecule studies of DNA mechanics. *Curr. Opin. Struct. Biol.* **2000**, *10*, 279–285. [CrossRef]
6. Forties, R.A.; Wang, M.D. Discovering the Power of Single Molecules. *Cell* **2014**, *157*, 4–7. [CrossRef]
7. Grier, D.G. A revolution in optical manipulation. *Nature* **2003**, *424*, 810–816. [CrossRef]
8. Ishii, Y.; Ishijima, A.; Yanagida, T. Single molecule nanomanipulation of biomolecules. *Trends Biotechnol.* **2001**, *19*, 211–216. [CrossRef]
9. Kim, H.; Ha, T. Single-molecule nanometry for biological physics. *Rep. Prog. Phys.* **2012**, *76*, 16601. [CrossRef]
10. Mehta, A.D.; Rief, M.; Spudich, J.A. Biomechanics, One molecule at a Time. *J. Biol. Chem.* **1999**, *274*, 14517–14520. [CrossRef]
11. Mehta, A.D.; Rief, M.; Spudich, J.A.; Smith, D.A.; Simmons, R.M. Single-molecule biomechanics with optical methods. *Science* **1999**, *283*, 1689–1695. [CrossRef] [PubMed]
12. Dickson, R.M.; Cubitt, A.B.; Tsien, R.Y.; Moerner, W.E. On/off blinking and switching behaviour of single molecules of green fluorescent protein. *Nature* **1997**, *388*, 355–358. [CrossRef] [PubMed]
13. Kellermayer, M.S.Z.; Smith, S.B.; Granzier, H.L.; Bustamante, C. Folding-Unfolding Transitions in Single Titin Molecules Characterized with Laser Tweezers. *Science* **1997**, *276*, 1112–1116. [CrossRef] [PubMed]
14. Rief, M.; Gautel, M.; Oesterhelt, F.; Fernandez, J.M.; Gaub, H.E. Reversible unfolding of individual Titin immunoglobulin domains by AFM. *Science* **1997**, *276*, 1109–1112. [CrossRef]
15. Tskhovrebova, L.; Trinick, J.; Sleep, J.A.; Simmons, R.M. Elasticity and unfolding of single molecules of the giant muscle protein titin. *Nature* **1997**, *387*, 308–312. [CrossRef]
16. Finer, J.T.; Simmons, R.M.; Spudich, J.A. Single myosin molecule mechanics: Piconewton forces and nanometre steps. *Nature* **1994**, *368*, 113–119. [CrossRef]

17. Molloy, J.E.; Burns, J.E.; Kendrick-Jones, J.; Tregear, R.T.; White, D.C.S. Movement and force produced by a single myosin head. *Nature* **1995**, *378*, 209–212. [CrossRef]

18. Svoboda, K.; Schmidt, C.F.; Schnapp, B.J.; Block, S.M. Direct observation of kinesin stepping by optical trapping interferometry. *Nature* **1993**, *365*, 721–727. [CrossRef]

19. Ashkin, A. Acceleration and trapping of particles by radiation pressure. *Phys. Rev. Lett.* **1970**, *24*, 156–159. [CrossRef]

20. Ashkin, A. Forces of a single-beam gradient laser trap on a dielectric sphere in the ray optics regime. *Biophys. J.* **1992**, *61*, 569–582. [CrossRef]

21. Svoboda, K.; Block, S.M. Biological applications of optical forces. *Annu. Rev. Biophys. Biomol. Struct.* **1994**, *23*, 247–285. [CrossRef] [PubMed]

22. Smith, S.B.; Cui, Y.; Bustamante, C. Overstretching B-DNA: The elastic response of individual double-stranded and single-stranded DNA molecules. *Science* **1996**, *271*, 795–799. [CrossRef] [PubMed]

23. Wang, M.D.; Yin, H.; Landick, R.; Gelles, J.; Block, S.M. Stretching DNA with optical tweezers. *Biophys. J.* **1997**, *72*, 1335–1346. [CrossRef]

24. Liphardt, J.; Onoa, B.; Smith, S.B.; Tinoco, I.J.; Bustamante, C. Reversible unfolding of single RNA molecules by mechanical force. *Science* **2001**, *292*, 733–737. [CrossRef] [PubMed]

25. Boer, G.; Johann, R.; Rohner, J.; Merenda, F.; Delacrétaz, G.; Renaud, P.; Salathé, R.-P. Combining multiple optical trapping with microflow manipulation for the rapid bioanalytics on microparticles in a chip. *Rev. Sci. Instrum.* **2007**, *78*, 116101. [CrossRef]

26. Gross, P.; Farge, G.; Peterman, E.J.G.; Wuite, G.J.L. Combining Optical Tweezers, Single-Molecule Fluorescence Microscopy, and Microfluidics for Studies of DNA–Protein Interactions. *Methods Enzymol.* **2010**, *475*, 427–453. [CrossRef]

27. Abrahamse, H.; Hamblin, M.R. New photosensitizers for photodynamic therapy. *Biochem. J.* **2016**, *473*, 347–364. [CrossRef]

28. Almaqwashi, A.A.; Zhou, W.; Naufer, M.N.; Riddell, I.A.; Yilmaz, Ö.H.; Lippard, S.J.; Williams, M.C. DNA intercalation facilitates efficient DNA-targeted covalent binding of phenanthriplatin. *J. Am. Chem. Soc.* **2019**, *141*, 1537–1545. [CrossRef]

29. Bahira, M.; McCauley, M.J.; Almaqwashi, A.A.; Lincoln, P.; Westerlund, F.; Rouzina, I.; Williams, M.C. A ruthenium dimer complex with a flexible linker slowly threads between DNA bases in two distinct steps. *Nucleic Acids Res.* **2015**, *43*, 8856–8867. [CrossRef]

30. Crisafuli, F.A.P.; Cesconetto, E.C.; Ramos, E.B.; Rocha, M.S. DNA–cisplatin interaction studied with single molecule stretching experiments. *Integr. Biol.* **2012**, *4*, 568–574. [CrossRef]

31. Sischka, A.; Toensing, K.; Eckel, R.; Wilking, S.D.; Sewald, N.; Ros, R.; Anselmetti, D. Molecular Mechanisms and Kinetics between DNA and DNA Binding Ligands. *Biophys. J.* **2005**, *88*, 404–411. [CrossRef] [PubMed]

32. Vladescu, I.D.; McCauley, M.J.; Nuñez, M.E.; Rouzina, I.; Williams, M.C. Quantifying force-dependent and zero-force DNA intercalation by single-molecule stretching. *Nat. Methods* **2007**, *4*, 517–522. [CrossRef] [PubMed]

33. Hembury, G.A.; Borovkov, V.V.; Inoue, Y. Chirality-sensing supramolecular systems. *Chem. Rev.* **2008**, *108*, 1–73. [CrossRef] [PubMed]

34. Kakiuchi, T.; Ito, F.; Nagamura, T. Time-resolved studies of energy transfer from meso-tetrakis(n-methylpyridinium-4-yl)-porphyrin to 3,3′-diethyl-2,2′-thiatricarbocyanine iodide along deoxyribonucleic acid chain. *J. Phys. Chem. B* **2008**, *112*, 3931–3937. [CrossRef] [PubMed]

35. Mutsamwira, S.; Ainscough, E.W.; Partridge, A.C.; Derrick, P.J.; Filichev, V.V. DNA duplex as a scaffold for a ground state complex formation between a zinc cationic porphyrin and phenylethynylpyren-1-yl. *J. Photochem. Photobiol. A Chem.* **2014**, *288*, 76–81. [CrossRef]

36. Pathak, P.; Yao, W.; Hook, K.D.; Vik, R.; Winnerdy, F.R.; Brown, J.Q.; Gibb, B.C.; Pursell, Z.F.; Phan, A.T.; Jayawickramarajah, J. Bright G-Quadruplex nanostructures functionalized with porphyrin lanterns. *J. Am. Chem. Soc.* **2019**, *141*, 12582–12591. [CrossRef]

37. Stulz, E. Nanoarchitectonics with porphyrin functionalized DNA. *Acc. Chem. Res.* **2017**, *50*, 823–831. [CrossRef]

38. Biebricher, A.S.; Heller, I.; Roijmans, R.F.H.; Hoekstra, T.P.; Peterman, E.J.G.; Wuite, G.J.L. The impact of DNA intercalators on DNA and DNA-processing enzymes elucidated through force-dependent binding kinetics. *Nat. Commun.* **2015**, *6*, 7304. [CrossRef]

39. Almaqwashi, A.A.; Paramanathan, T.; Rouzina, I.; Williams, M.C. Mechanisms of small molecule–DNA interactions probed by single-molecule force spectroscopy. *Nucleic Acids Res.* **2016**, *44*, 3971–3988. [CrossRef]
40. Fiel, R.J. Porphyrin—Nucleic acid interactions: A Review. *J. Biomol. Struct. Dyn.* **1989**, *6*, 1259–1274. [CrossRef]
41. Sehlstedt, U.; Kim, S.K.; Carter, P.; Goodisman, J.; Vollano, J.F.; Norden, B.; Dabrowiak, J.C. Interaction of cationic porphyrins with DNA. *Biochemistry* **1994**, *33*, 417–426. [CrossRef] [PubMed]
42. Chung, W.-J.; Cui, Y.; Chen, C.-S.; Wei, W.H.; Chang, R.-S.; Shu, W.-Y.; Hsu, I.C. Freezing shortens the lifetime of DNA molecules under tension. *J. Biol. Phys.* **2017**, *43*, 511–524. [CrossRef] [PubMed]
43. Cheng, W.L.; Erbay, C.; Sadr, R.; Han, A. Dynamic flow characteristics and design principles of laminar flow microbial fuel cells. *Micromachines* **2018**, *9*, 479. [CrossRef] [PubMed]
44. Bokstein, B.S.; Mendelev, M.I.; Srolovitz, D.J. *Thermodynamics and Kinetics in Materials Science: A Short Course*; Oxford University Press: New York, NY, USA, 2005; ISBN 9780198528036.
45. Sørensen, K.T.; Kristensen, A. Label-free monitoring of diffusion in microfluidics. *Micromachines* **2017**, *8*, 329. [CrossRef]
46. Eriksson, E.; Enger, J.; Nordlander, B.; Erjavec, N.; Ramser, K.; Goksör, M.; Hohmann, S.; Nyström, T.; Hanstorp, D. A microfluidic system in combination with optical tweezers for analyzing rapid and reversible cytological alterations in single cells upon environmental changes. *Lab Chip* **2007**, *7*, 71–76. [CrossRef]
47. Eriksson, E.; Scrimgeour, J.; Granéli, A.; Ramser, K.; Wellander, R.; Enger, J.; Hanstorp, D.; Goksör, M. Optical manipulation and microfluidics for studies of single cell dynamics. *J. Opt. A Pure Appl. Opt.* **2007**, *9*, S113. [CrossRef]
48. Eriksson, E.; Sott, K.; Lundqvist, F.; Sveningsson, M.; Scrimgeour, J.; Hanstorp, D.; Goksör, M.; Granéli, A. A microfluidic device for reversible environmental changes around single cells using optical tweezers for cell selection and positioning. *Lab Chip* **2010**, *10*, 617–625. [CrossRef]
49. Yan, X.; Habbersett, R.C.; Cordek, J.M.; Nolan, J.P.; Yoshida, T.M.; Jett, J.H.; Marrone, B.L. Development of a mechanism-based, DNA staining protocol using SYTOX orange nucleic acid stain and DNA fragment sizing flow cytometry. *Anal. Biochem.* **2000**, *286*, 138–148. [CrossRef]
50. Kellermayer, M.S.Z.; Smith, S.B.; Bustamante, C.; Granzier, H.L. Complete unfolding of the titin molecule under external force. *J. Struct. Biol.* **1998**, *122*, 197–205. [CrossRef]
51. Mandal, S.S.; Merz, D.R.; Buchsteiner, M.; Dima, R.I.; Rief, M.; Žoldák, G. Nanomechanics of the substrate binding domain of Hsp70 determine its allosteric ATP-induced conformational change. *Proc. Natl. Acad. Sci. USA* **2017**, *114*, 6040–6045. [CrossRef]
52. Caldarini, M.; Sonar, P.; Valpapuram, I.; Tavella, D.; Volonté, C.; Pandini, V.; Vanoni, M.A.; Aliverti, A.; Broglia, R.A.; Tiana, G.; et al. The complex folding behavior of HIV-1-protease monomer revealed by optical-tweezer single-molecule experiments and molecular dynamics simulations. *Biophys. Chem.* **2014**, *195*, 32–42. [CrossRef] [PubMed]
53. Hao, Y.; Canavan, C.; Taylor, S.S.; Maillard, R.A. Integrated method to attach DNA handles and functionally select proteins to study folding and protein-ligand interactions with optical tweezers. *Sci. Rep.* **2017**, *7*, 10843. [CrossRef] [PubMed]
54. Yu, Z.; Cui, Y.; Selvam, S.; Ghimire, C.; Mao, H. Dissecting cooperative communications in a protein with a high-throughput single-molecule scalpel. *ChemPhysChem* **2015**, *16*, 223–232. [CrossRef]

 *micromachines*

Article

Optical Trapping and Manipulating with a Silica Microring Resonator in a Self-Locked Scheme

Victor W. L. Ho [1,†], Yao Chang [2,†], Yang Liu [2], Chi Zhang [2], Yuhua Li [1], Roy R. Davidson [3], Brent E. Little [4], Guanghui Wang [2,*] and Sai T. Chu [1,*]

1   Department of Physics, City University of Hong Kong, Kowloon 999077, Hong Kong, China; wailokho2-c@my.cityu.edu.hk (V.W.L.H.); yuhuali3-c@my.cityu.edu.hk (Y.L.)

2   College of Engineering and Applied Sciences, Nanjing University, Nanjing 210093, China; 131270047cy@sina.com (Y.C.); MF1934012@smail.nju.edu.cn (Y.L.); MF1834026@smail.nju.edu.cn (C.Z.)

3   QXP Technology, Xi'an 710311, China; roy.davidson@qxptech.com

4   State Key Laboratory of Transient Optics and Photonic, Xi'an Institute of Optics and Precision Mechanics, Chinese Academy of Sciences, Xi'an 710119, China; brent.little@opt.ac.cn

*   Correspondence: wangguanghui@nju.edu.cn (G.W.); saitchu@cityu.edu.hk (S.T.C.); Tel.: +86-025-83593302 (G.W.); +852-3442-4968 (S.T.C.)

†   Those authors equally contributed to this work.

Received: 31 December 2019; Accepted: 13 February 2020; Published: 15 February 2020

Abstract: Based on the gradient force of evanescent waves in silica waveguides and add-drop micro-ring resonators, the optical trapping and manipulation of micro size particles is demonstrated in a self-locked scheme that maintains the on-resonance system even if there is a change in the ambient temperature or environment. The proposed configuration allows the trapping of particles in the high Q resonator without the need for a precise wavelength adjustment of the input signal. On the one hand, a silicon dioxide waveguide having a lower refractive index and relatively larger dimensions facilitates the coupling of the laser with a single-mode fiber. Furthermore, the experimental design of the self-locked scheme reduces the sensitivity of the ring to the environment. This combination can trap the micro size particles with a high stability while manipulating them with high accuracy.

Keywords: optical trapping; microring resonator; micro manipulation

---

1. Introduction

Optical tweezers are widely used for capturing micro-particles or cells, due to the strong gradient force of highly focused laser beams [1,2]. To improve the level of system integration and trapping stiffness, researchers have explored near-field optical forces of evanescent fields around photonic waveguides in recent years [3–7]. Several methods for particles trapping and manipulation have been demonstrated, such as the optofluidic switching of nanoparticles [8,9], nanoparticle sorting [10,11], storing of nanoparticles [12,13], and precise manipulation in nanophotonic Standing Wave Array Traps (nSWATs) [14–16], among which the micro-ring resonator plays an important role. Various demonstrations of optical trapping with planar microring resonators have previously been reported, e.g., the authors of [17] studied particles that are trapped and manipulated by adjusting the input tunable lasers both on and off resonance. A resonator is an energy storage system with a feedback system where the optical intensity in the cavity is enhanced when it is on resonance. Compared to the waveguide trapping configurations, resonant cavity trapping presents two advantages [17]; the strong force enhancement due to the high field confinement in a cavity, and the cavity perturbation induced by the trapped object, which could serve as a highly sensitive probe for analyzing the physical properties of the objects [18,19]. The trapping and manipulation of micro particles on planar ring resonators have recently been demonstrated using silicon photonic crystal resonators [20]. However, to be able to trap the particles continually without interruption, it is vitally important to keep the input laser frequency aligned with the resonance frequency of the resonator, as the resonant frequency may shift during the operation from any changes in the ambient temperature and environment. The misalignment of the laser with the resonance frequencies reduces the intracavity power of the resonator, causing the particle to dislodge from the trap, resulting in the loss of control of the trapped particle. [21]. This is especially important when applying optical trapping with high-Q ring resonators in highly precise measurements [17] and in biomedical [22,23], biochemical, and chemical sensing [24]. To improve the stability of the trapping process, we proposed a self-locked scheme which allows the signal to remain on resonance regardless of the variation in the ambient temperature and environment [25].

The proposed self-locked configuration is shown in Figure 1a, where we replace the tunable laser in the conventional open-loop configuration with a bandpass filter and arrange the circuit in a closed-loop configuration. The configuration is similar to the passive mode-locked or fiber laser configuration where the external cavity acts as the gain cavity and the combination of the bandpass filter with the ring resonator acts as the wavelength selective element [21]. Since the external cavity has a much finer free-spectral range (FSR) spacing and narrower linewidth than the ring resonator, the system remains in the lasing condition regardless of any resonance frequency shift encountered by the ring resonator. Additional control elements, such as switches, can be added to the basic configuration in Figure 1a to select the propagation direction of the particles, while the propagation speed can be adjusted from the power of the amplifier. When it is on resonance, the intracavity power in the microring resonator is proportional to its Q factor, with $Q = \lambda/\Delta\lambda$, where $\Delta\lambda$ is the full-width half maximum (FWHM) of the resonance peak. As the light resonates and builds up within the resonator, the intracavity power and field in the microring resonator can be enhanced by orders of magnitude [25] above the input power. Thus, the field in the ring can be much higher than the bus waveguide, and trapping can be achieved with a much lower input power as compared to particle trapping with a simple waveguide. In this work, we demonstrate that the proposed scheme can trap polystyrene (PS) beads of sizes up to 3 µm with an average input power as low as a few tens of mW.

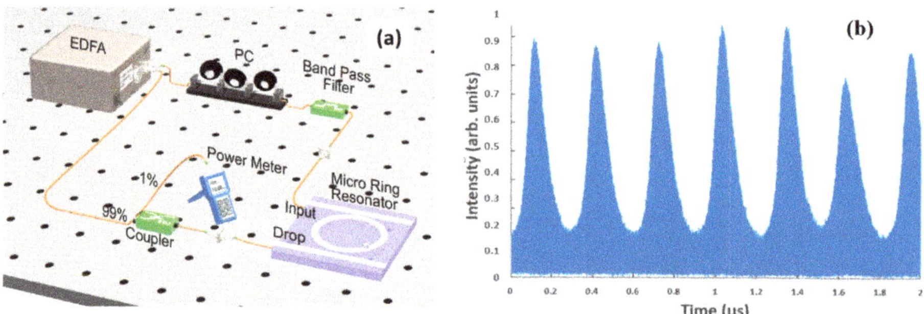

Figure 1. (**a**) The optical trapping system with a highly-doped silica glass ring resonator in a self-locked configuration, EDFA is an erbium-doped fiber amplifier, and PC is a three-ring polarization controller. (**b**) The fast detector output from the tap showing Q-switching in the fiber loop cavity. The repetition rate of the Q-switched pulses is approximately 100 kHz with the high frequency component within the Q-switched pulses at 6.6 MHz, which corresponds to a fiber loop length of 30 m.

2. Materials and Methods

The self-locked optical trapping setup used in the experiment in Figure 1a consists of an amplifying fiber ring loop of approximately 30 m [21] corresponding to an FSR of approximately 6.6 MHz. The FSR of the external cavity loop represents the resolution of the measurement system, while the role of the ring resonator is to select which of the resonance frequencies of the external loop will be amplified. By choosing the bandpass filter to be larger than the FSR of the ring resonator, one ensures that at least one of the ring resonances is excited regardless of any resonance frequency drift. However, to prevent mode hopping between different resonances, the bandpass filter is selected to be close to the FSR of the ring resonator. Since the fiber loop of the main cavity is relatively long, one does not expect continuous wave (CW) lasing from the configuration due to modulation instability. To investigate the dynamic behavior of the optical signal in the fiber loop, we measured the time response of the optical signal at the tap with a high-speed optical detector. The output waveform from the detector is shown in Figure 1b, which shows an irregular pattern of Q switch pulses due to the dynamic instability in the long cavity loop.

The investigation of the trapping capability with integrated optical waveguides and with the enhancement of the light intensity in ring resonators has been demonstrated in silicon-rich integrated optical circuits, such as silicon nitride waveguides [26]. The ring resonator add/drop filter used in this experiment is fabricated with a low-loss highly doped glass with a core index of 1.70, a cladding index of 1.445, and with a waveguide cross section of 1.45 μm $\times$ 1.45 μm to allow a higher field penetration into the cladding for trapping while maintaining a reasonable Q factor. With the ring radius at 47 μm, it corresponds to an FSR of 569 GHz. The gap separation between the ring and the bus waveguide is 0.85 μm, so that the FWHM of the resonance peak is 21 pm at around 1550 nm. This corresponds to a Q factor of approximately 75,000 when the device is measured in air but the FWHM, Q factor, and FSR change to 27 pm, 65,000 and 572 GHz, respectively, when it is immersed in water. It is important to note that the trapping force is a function of the intensity enhancement factor B of the resonator, which measures the amount of amplification of the input in the resonator cavity. When the device is placed in air the resonator enhances the intensity by approximately 50 times, but this is reduced to only 30 times when it is in water due to the weaker field confinement. Furthermore, to provide a flat surface for the trapping application, the device was planarized to the top of the waveguide. The device was pigtailed to an optical fiber array with a coupling loss of approximately 5 dB/facet. In the experiment, we first deposited deionized water droplets on the surface of the chip and placed the PS particles in the vicinity of the waveguide structures. A thin cover glass was then placed on the water surface of the chip to provide an even distribution of the water solution across the surface and to allow the experiment to be monitored via a microscope and Charge Coupled Devices (CCD) camera system.

3. Simulation

To investigate the interaction between the light field on the waveguide and the particle, we calculated the distribution of the light field in both the Transverse Magnetic (TM) mode and Transverse Electric (TE) mode of the waveguide cross section. The light transmission is in the x-direction, the vertical direction of the cross section is the z-direction, and the horizontal direction is the y-direction. As shown in Figure 2a,b, the evanescent field in the TM mode was stronger in the z-direction, and that in the TE mode was stronger in the y-direction. Based on the requirement of trapping particles at the waveguide top surface, the TM mode was adopted for subsequent simulations. The simulations were carried out using the commercial software packages from Lumerical, such as FDTD and MODE Solutions, with the finite-difference time-domain algorithm to solve the electromagnetic field distribution, while the volumetric technique algorithm was used to calculate the optical force on the particles. The volumetric technique algorithm uses the Coulomb force formula, the Lorentz formula, and the Maxwell's equations to solve the optical force at a point on a particle, and the optical force on each point is integrated over the entire particle volume to obtain the optical force on the particle. Compared with a silicon waveguide, our cross-section mode was larger and had the advantage of attaching an optical fiber, which allowed the device to be portable for the ease of the setup. Both the TM and TE mode had relatively small losses. It can be seen from the simulation that the TM mode played a major role in capturing particles on the upper surface of the waveguide.

We also simulated the scattering force along the x-direction and the gradient force along the y-direction on a particle with a diameter of 3 μm. The scattering force was stable at around 0.01 pN/mW, which formed the driving force for the particles to advance along in the x-direction. The particle velocity of the simulation was calculated by Stokes' drag formula:

$$v = F/3\pi D\eta \tag{1}$$

where Fs is the scattering force, D is the particle diameter, and η is the viscosity coefficient.

As shown in Figure 2c, the simulation and experimental results were in good agreement within a certain power range. When the gradient force was integrated along the y-direction displacement, the corresponding optical potential well could be obtained. In the case of 20 mW, the depth of the potential well in Figure 2d was around 80 k_BT, when the polystyrene particles with a diameter of 3 µm were used at a speed of 100 µm/s, according to the formula of centrifugal force:

$$F = mv^2/r \tag{2}$$

where F is the centrifugal force, m is the mass of the particle, and v is the viscosity. The required centripetal force is about 0.1 fN. In the simulation, when the incident light power was 20 mW, the maximum gradient force on the particle was 0.3 fN, so the gradient force could provide the centripetal force required for the particle to move around the ring.

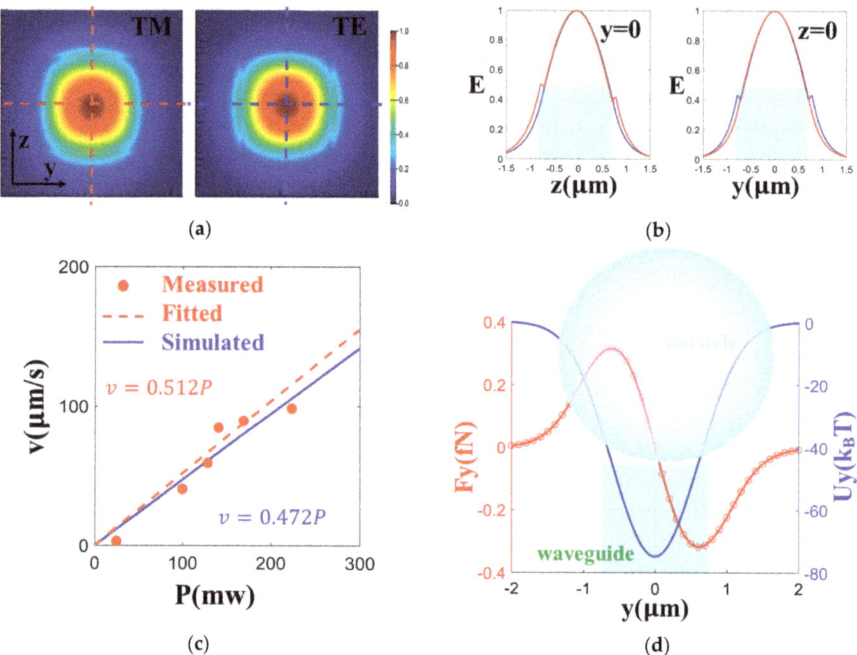

Figure 2. (**a**) The TM and TE mode distribution in the waveguide cross section. (**b**) The electric field *E* of four dotted lines passing through the centre of the section with a change with the z-direction (left) or y-direction (right). The line in TM is blue, and the line in TE is red. (**c**) The fitting red dotted line between the particle velocity *v* and power *P* obtained by the data in the experiment (red dots). The corresponding simulation results are represented with a blue line. (**d**) The gradient force and the corresponding optical potential well along the y-direction.

We also simulated the light field of the whole structure and plotted the transmission spectra of the rings at the drop port, as shown in Figure 3. It was determined to have an FSR of 4.7 nm with an FWHM of 24 pm at around 1564 nm, with a Q factor of 65,154. These results are consistent with the experimental results shown in the next section. The calculated Q factor at the resonance wavelength (λ) was compared with the experiment to verify the simulation.

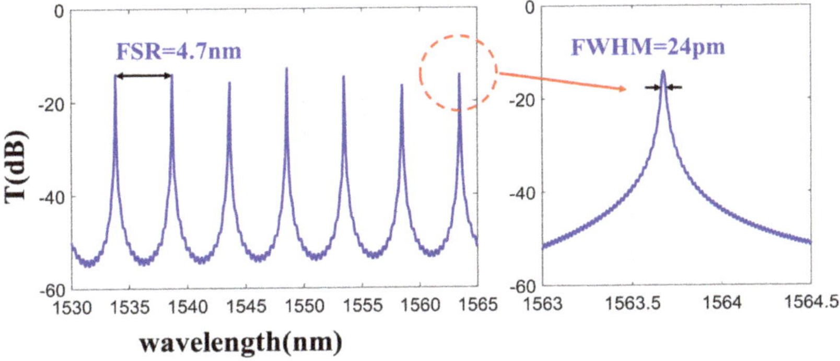

Figure 3. The simulated transmission spectra of rings at the drop from 1530 nm to 1565 nm.

Using the different enhancement factors B, we calculated a series of potential wells along the y-direction, which reflected the process of particle transfer from bus waveguide to ring. In order to show the process better, we calculated the particle motion considering the Brownian motion. The relationship between the displacement of the particle and time was obtained by solving the Newton equation of motion for a single particle, including the optical gradient force, the viscous resistance, and the Brownian force [27]. According to the displacement statistics of 1000 particles in 5 s, the distribution ratio of particles on the bus waveguide and ring was obtained, as shown in Figure 4. The simulation time was set to 5 s, because the gap between the ring and the bus waveguide changes little during this time, taking into account the particle actual x-direction velocity [12]. With the increase of B, the proportion of particles jumping from the bus to ring increased, as shown in Figure 4b–d. This is a good explanation for the phenomenon of particles transferring from bus waveguide to ring.

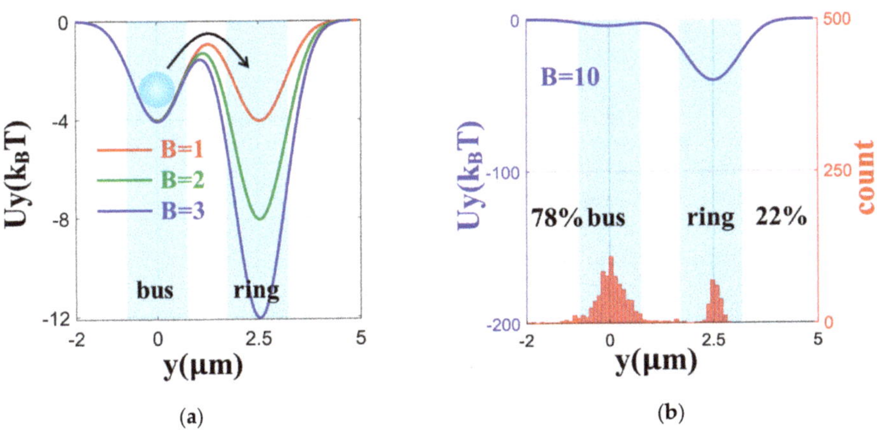

Figure 4. *Cont.*

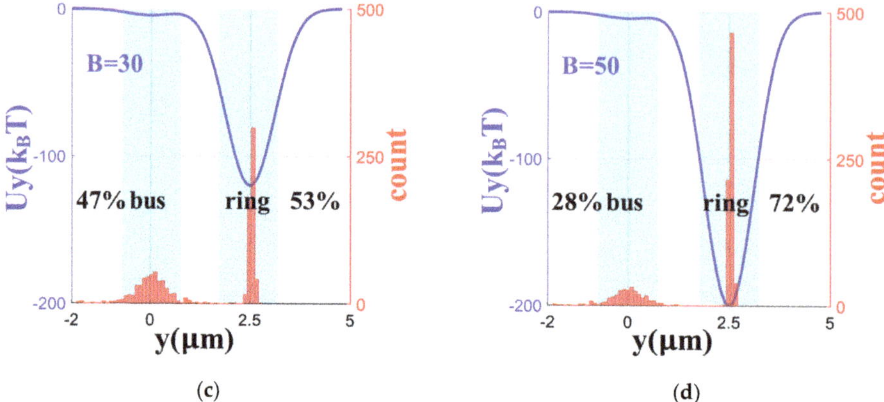

Figure 4. (**a**) Potential wells of 3 μm diameter spheres in the y-direction with B = 1, 2, and 3. (**b**) The particle distribution on the ring and bus in 5 s when B = 10. The particle distribution on the ring and bus in 5 s when (**c**) B = 30 and (**d**) B = 50.

4. Results and Discussion

Q-switched pulses with a high frequency component of ~6.6 MHz equal to the free spectral range of the external cavity loop similar to those shown in Figure 1b were observed in this experiment. Stable periodic pulses can also be observed when the external cavity length matches integer multiples of the microring resonator cavity length, but due to the long cavity length stable passive mode locking is not sustainable. Methods such as adding modulation or drastically reducing the length of the external loop will be needed to maintain stable mode locking. Nevertheless, the self-locked configuration allows the microring resonator to always remain on-resonance, producing the maximum trapping force on the resonator.

In the experiment, the particles and deionized water solution were first dropped onto the top surface of the chip. Since there is no fluidic cell to guide the particles, and in view of the flat chip surface, the particles flowed freely on the chip surface when the EDFA was off. When the power of the amplifier was switched on, power was delivered to the bus waveguide, and the intensity in the ring resonator started to build up. At this point, the particles that were in the vicinity of the waveguides started to migrate toward the waveguides due to the gradient force produced by the intensity in the waveguide. When the particles were close enough to the waveguide, they would be trapped by the waveguides and pushed along the waveguide by the scattering force, as shown in Figure 5a. For those particles trapped by the input waveguides, they would be pushed along the waveguide until they reached the ring resonator where they could be attracted and trapped by the ring resonator because of the higher intensity inside the ring cavity.

Since the intensities in the bus and ring waveguides are not the same, the particle encounters different scattering forces when it is on the bus and on the ring waveguides. The ratio between the speeds of the particle when it is on the bus waveguide and when it is on the ring resonator provides a direct measure of the intensity enhancement factor of the ring resonator. Figure 5a and Video S1 show two beads that were initially traveling along the bus waveguide and were later trapped by the ring resonator when they arrived at the coupling region between the ring and the bus. From the two measured speeds, we found that the speed of the particle was approximately 15 times faster when it was on the ring resonator, which corresponded to the intensity differences between the ring and the bus. Figure 5b and Video S2 shows the two different force observed in the experiment: the scattering force (Fs) that pushed the group of beads along the ring resonator in the direction indicated by the red arrow, and the gradient force (Fg) that pulled the free-floating particle toward the waveguide indicated by the bluearrow.

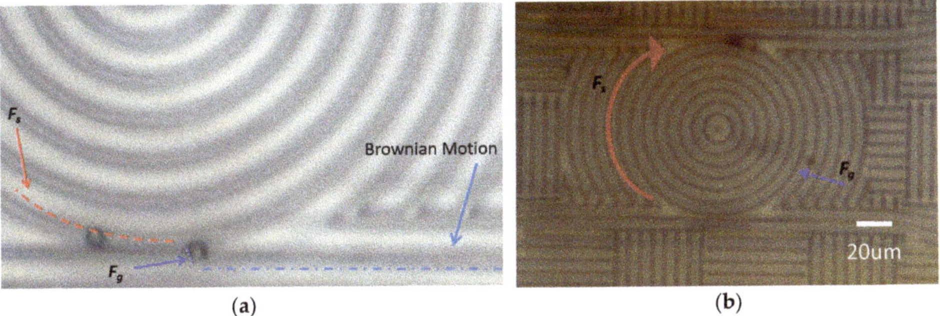

(**a**)   (**b**)

Figure 5. (**a**) The optical image of the ring resonator (highlighted by the red dotted line with the ring and the blue dotted line with the bus) prior to the EDFA being switched on. This figure shows two beads delivered along the bus waveguide to the ring. The average speed of a bead on the bus and the ring is 3.13 µm/s and 40.38 µm/s, and the electrical field of the ring is stronger than the bus waveguide. (**b**) The demonstration of the two trapping forces produced by the intensity in the waveguide. The four beads are being pushed around the ring resonator by the scattering force F_s indicated by the red arrow, and the single bead outside the ring is being pulled toward the waveguide by the gradient force F_g, indicated by the blue arrow.

We also observed a series of three particles that were trapped and pushed along the ring resonator at the same time while a single particle was trapped by the same ring, as shown in Figure 6a. In the example shown in Figure 6, the four particles were initially separated into two groups, a single particle and a group of three particles. The two groups of particles travelled at different speeds, with the group of three travelling at a higher speed than the single particle. The measured speed of the single particle was 19.7 µm/s, while the speed of the peloton of three particles was 23.6 µm/s. Figure 6a,b are still images of the Video S3 taken at 30 s apart. Initially, the single particle was approximately one quarter of the ring circumference ahead of the peloton, but at 30 s the peloton caught up with the single particle due to the reduced aerodynamic drag [28]. It is also interesting to note that all three particles received the same amount of force due to the same field overlap they received, but they shared the burden to overcome the drag.

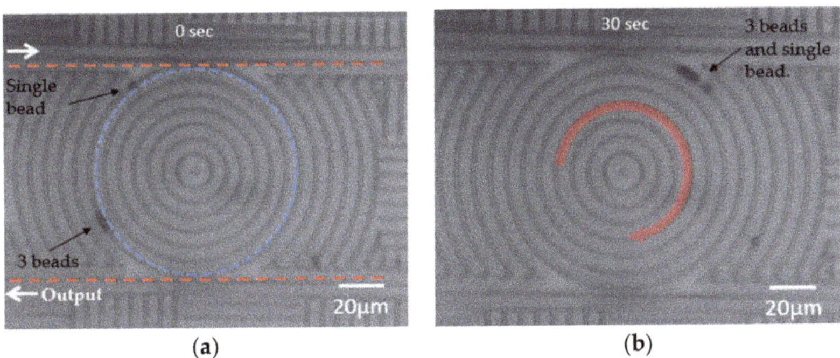

(**a**)   (**b**)

Figure 6. Four particles trapped on top of the ring waveguide in two groups traveling at different speeds. (**a**) Initially, they were separated by approximately one quarter of the ring circumference, with the group of three beads traveling at a higher speed. (**b**) The group of three beads caught up with the single bead after 30 s.

The speed of the particle can also be adjusted by controlling the power of the EDFA; however, the amplitude and repetition rate of the Q-switched pulse are highly sensitive to changes in the polarization and cavity length variation. Hence, we can only relate the average power in the resonator to the trapping speed, instead of the peak pulse power. Table 1 shows the resonance wavelength and measured speed of the 3 μm size PS particle at the different average powers with the particle on the waveguide and on the resonator. At the average input power of 14.1 dBm, the speed of the particle was measured at 3 μm/s, however the speed increased to 40.4 μm/s when it was transferred onto the ring resonator. Here, one can roughly adjust the speed of the particle on the ring resonator from 40.4 μm/s to 98 μm/s by changing the input average power from 14.1 dBm to 17.5 dBm. Figure 7a shows the measured FWHM of the resonance peak as a function of the wavelength when the device is in air and when it is immersed in water. For the ring resonator used in the experiment, the FWHM of the resonance peak increased by approximately 50% when the device was immersed in water as compared to when it was in the air. Besides increasing the FWHM, the higher refractive index and absorption of water also decreased the field enhancement factor B, as shown in Figure 7b.

Finally, in the experiments described in this section, the device was exposed to the ambient environment without any mechanism to control the device temperature or resonance location; therefore, the resonance frequency was allowed to drift. Although the resonance frequency can drift on the order of tens of pm due to the intensity induced temperature change, as shown in Table 1, the self-locked scheme was able to keep the system on resonance, maintaining the trapping capability at all times.

Table 1. The resonance wavelength and estimated average power in the waveguide and the resonator with the relative trapping speed.

Resonance Wavelength (nm)	Average Power in the Input Waveguide (dBm)	Average Power in the Resonator (dBm)	Location of the Particle	Trapping Speed (μm/s)
1555.35	14.1	N/A	Bus	3.13
1555.35	14.1	20.0	Ring	40.4
1555.38	15.1	21.1	Ring	59.1
1555.39	15.5	21.5	Ring	84.4
1555.40	16.3	22.3	Ring	89.5
1555.41	17.5	23.5	Ring	98.4

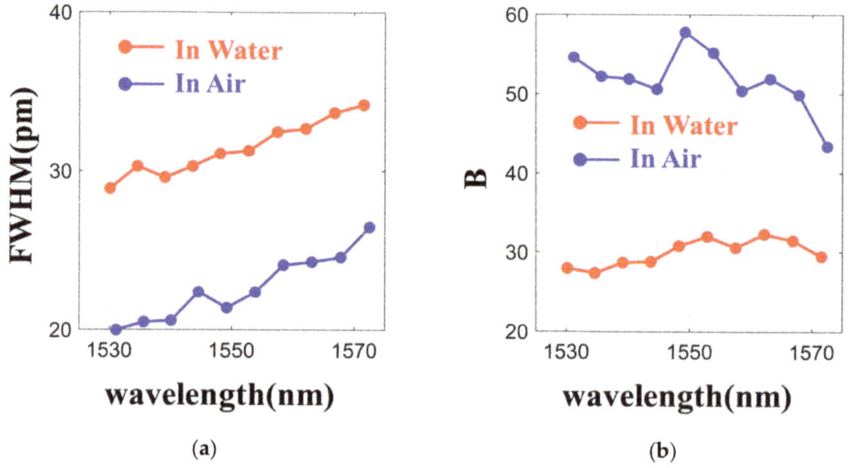

Figure 7. (**a**) The FWHM in different wavelengths in air and water. (**b**) The confinement factor (B) in different wavelengths in air and in water.

5. Conclusions

We have proposed and demonstrated an optical trapping scheme that allows the continuous trapping and manipulation of polystyrene beads, with sizes as large as 3 μm in diameter, on an integrated optical microring resonator. The proposed scheme employs a closed-loop configuration that can keep the microring resonator on-resonance at all times, trapping the particles on top of the resonator regardless of how the particles affect the resonance frequency of the resonator. Compared to the conventional trapping configuration, the proposed system eliminates the need for the *in-situ* adjustment of an input wavelength to align with the ring resonance. We have carried out a detailed simulation of the trapping process that generally agrees with the experimental results. We have shown that the speed of the particles can be adjusted by the EDFA power level, where the particle can be propelled around the ring resonator with speeds close to 100 μm/s. It is important to point out that the signal that travels around the resonator is in the form of pulses, where each of the pulses contains the instantaneous time response of the environment of the resonator. One can explore the characteristics of these pulses to provide real-time information of the particles on the resonator in highly sensitive sensing systems.

Supplementary Materials: The following are available online at http://www.mdpi.com/2072-666X/11/2/202/s1, Video S1: Refer to Figure 5a, Video S2: Refer to Figure 5b, Video S3: Refer to Figure 6a,b.

Author Contributions: Conceptualization, S.T.C. and G.W.; methodology, V.W.L.H. and Y.C.; software, Y.L. (Yang Liu); validation, V.W.L.H.,Y.C. and Y.L. (Yang Liu); formal analysis, Y.C.; investigation, V.W.L.H., Y.L. (Yuhua Li); resources, S.T.C.; fabrication, R.R.D.; design the device, B.E.L.; data curation, Y.C.; writing—original draft preparation, V.W.L.H.; writing—review and editing, G.W.; visualization, C.Z.; supervision, S.T.C.; project administration, G.W.; funding acquisition, G.W. All authors have read and agreed to the published version of the manuscript.

Funding: This work was funded by National Natural Science Foundation of China (nos. 61875083, 61535005); Social Development Project of Jiangsu Province (BE2019761); Research Grants Council, University Grants Committee (GRF 11213618); and Strategic Priority Research Program of the Chinese Academy of Sciences (XDB24030300).

Acknowledgments: The authors thank Feng Wang and Tianying Sun for help with equipment and the materials support.

Conflicts of Interest: The authors declare no conflict of interest.

References

1. Ashkin, A.; Dziedzic, J.M.; Yaman, T. Optical trapping and manipulation of single cells using infrared laser beams. *Nature* **1987**, *330*, 769–771. [CrossRef] [PubMed]
2. Wang, M.M.; Tu, E.; Raymond, D.E.; Yang, J.M.; Zhang, H.C.; Hagen, N.; Dees, B.; Mercer, E.M.; Forster, A.H.; Kariv, I.; et al. Microfluidic sorting of mammalian cells by optical force switching. *Nat. Biotechnol.* **2005**, *23*, 83–87. [CrossRef] [PubMed]
3. Schmidt, B.S.; Yang, A.H.J.; Erickson, D.; Lipson, M. Optofluidic trapping and transport on solid core waveguides within a microfluidic device. *Opt. Express* **2007**, *15*, 14322–14334. [CrossRef]
4. Yang, A.H.J.; Lerdsuchatawanich, T.; Erickson, D. Forces and Transport Velocities for a Particle in a Slot Waveguide. *Nano Lett.* **2009**, *9*, 1182–1188. [CrossRef] [PubMed]
5. Cai, H.; Poon, A.W. Optical manipulation and transport of microparticle on silicon nitride microring resonator—Based add-drop devices. *Opt. Lett.* **2010**, *35*, 2855–2857. [CrossRef] [PubMed]
6. Chen, Y.F.; Serey, X.; Sarkar, R.; Chen, P.; Erickson, D. Controlled Photonic Manipulation of Proteins and Other Nanomaterials. *Nano Lett.* **2012**, *12*, 1633–1637. [CrossRef]
7. Soltani, M.; Lin, J.; Forties, R.A.; Inman, J.T.; Saraf, S.N.; Fulbright, R.M.; Lipson, M.; Wang, M.D. Nanophotonic trapping for precise manipulation of biomolecular arrays. *Nat. Nanotechnol.* **2014**, *9*, 448–452. [CrossRef]
8. Jiao, W.X.; Wang, G.H.; Ying, Z.F.; Zou, Y.; Ho, H.P.; Sun, T.Y.; Huang, Y.; Zhang, X.P. Switching of nanoparticles in large-scale hybrid electro-optofluidics integration. *Opt. Lett.* **2016**, *41*, 2652–2655. [CrossRef]
9. Jiao, W.X.; Wang, G.H.; Ying, Z.F.; Kang, Z.W.; Sun, T.Y.; Zou, N.M.; Ho, H.P.; Zhang, X.P. Optofluidic Switching of Nanoparticles Based on a WDM Tree Splitter. *IEEE Photonics J.* **2016**, *8*. [CrossRef]

10. Xu, X.F.; Dong, Y.M.; Wang, G.H.; Jiao, W.X.; Ying, Z.F.; Ho, H.P.; Zhang, X.P. Reconfigurable Sorting of Nanoparticles on a Thermal Tuning Silicon Based Optofluidic Chip. *IEEE Photonics J.* **2017**, *10*. [CrossRef]

11. Xu, X.F.; Wang, G.H.; Jiao, W.X.; Ji, W.B.; Jiang, M.; Zhang, X.P. Multi-level sorting of nanoparticles on multi-step optical waveguide splitter. *Opt. Express* **2018**, *26*, 29262–29271. [CrossRef] [PubMed]

12. Xu, W.H.; Wang, Y.Y.; Jiao, W.X.; Wang, F.; Xu, X.F.; Jiang, M.; Ho, H.P.; Wang, G.H. Tunable optofluidic sorting and manipulation on micro-ring resonators from a statistics perspective. *Opt. Lett.* **2019**, *44*, 3226–3229. [CrossRef] [PubMed]

13. An, R.; Wang, G.H.; Ji, W.B.; Jiao, W.X.; Jiang, M.; Chang, Y.; Xu, X.F.; Zou, N.M.; Zhang, X.P. Controllable trapping and releasing of nanoparticles by a standing wave on optical waveguides. *Opt. Lett.* **2018**, *43*, 3901–3904. [CrossRef] [PubMed]

14. Ye, F.; Badman, R.P.; Inman, J.T.; Soltani, M.; Killian, J.L.; Wang, M.D. Biocompatible and High. Stiffness Nanophotonic Trap Array for Precise and Versatile Manipulation. *Nano Lett.* **2016**, *16*, 6661–6667. [CrossRef]

15. Ye, F.; Soltani, M.; Inman, J.T.; Wang, M.D. Tunable nanophotonic array traps with enhanced force and stability. *Opt. Express* **2007**, *25*, 7907–7918. [CrossRef]

16. Badman, R.P.; Ye, F.; Caravan, W.; Wang, M.D. High Trap Stiffness Microcylinders for Nanophotonic Trapping. *ACS Appl. Mater. Interfaces* **2019**, *11*, 25074–25080. [CrossRef]

17. Lin, S.; Schonbrun, E.; Crozier, K. Optical Manipulation with Planar Silicon Microring Resonators. *Nano Lett.* **2010**, *10*, 2408–2411. [CrossRef]

18. Vollmer, F.; Arnold, S. Whispering-gallery-mode biosensing: Label-free detection down to single molecules. *Nat. Methods* **2008**, *5*, 591–596. [CrossRef]

19. Zhu, J.; Ozdemir, S.K.; Xiao, Y.F.; Li, L.; He, L.; Chen, D.R.; Yang, L. On-chip single nanoparticle detection and sizing by mode splitting in an ultrahigh-Q microresonator. *Nat. Photonics* **2010**, *4*, 46–49. [CrossRef]

20. Mandal, S.; Serey, X.; Erickson, D. Nanomanipulation Using Silicon Photonic Crystal Resonators. *Nano Lett.* **2010**, *10*, 99–104. [CrossRef]

21. Peccianti, M.; Pasquazi, A.; Park, Y.; Little, B.E.; Chu, S.T.; Moss, D.J.; Morandotti, R. Demonstration of a stable ultrafast laser based on a nonlinear microcavity. *Nat. Commun.* **2012**, *3*, 765. [CrossRef] [PubMed]

22. Baker, J.E.; Badman, R.P.; Wang, M.D. Nanophotonic trapping: Precise manipulation and measurement of biomolecular arrays. *WIREs Nanomed. Nanobiotechnol.* **2018**, *10*, e1477. [CrossRef] [PubMed]

23. Badman, R.P.; Ye, F.; Wang, M.D. Towards biological applications of nanophotonic tweezers. *Curr. Opin. Chem. Biol.* **2019**, *53*, 158–166. [CrossRef] [PubMed]

24. Sun, Y.; Fan, X. Optical ring resonators for biochemical and chemical sensing. *Anal. Bioanal. Chem.* **2011**, *399*, 205–211. [CrossRef]

25. Little, B.E.; Chu, S.T.; Haus, H.A.; Foresi, J.; Laine, J.P. Microring resonator channel dropping filters. *J. Lightwave Technol.* **1997**, *15*, 998–1005. [CrossRef]

26. Wang, H.; Wu, X.; Shen, D. Localized optical manipulation in optical ring resonators. *Opt. Express* **2015**, *23*, 27650–27660. [CrossRef]

27. Michaelides, E.E. Brownian movement and thermophoresis of nanoparticles in liquids. *Int. J. Heat Mass Transf.* **2015**, *81*, 179–187. [CrossRef]

28. Blocken, B.; van Druenen, T.; Toparlar, Y.; Malizia, F.; Mannion, P.; Andrianne, T.; Marchal, T.; Maas, G.J.; Diepens, J. Aerodynamic drag in cycling pelotons: New insights by CFD simulation and wind tunnel testing. *J. Wind Eng. Ind. Aerodyn.* **2018**, *179*, 319–337. [CrossRef]

 *micromachines*

Communication

Auxiliary Optomechanical Tools for 3D Cell Manipulation

Ivan Shishkin [1,2,3,*], Hen Markovich [2,3], Yael Roichman [3,4,5] and Pavel Ginzburg [2,3]

1   Faculty of Physics and Engineering, ITMO University, Lomonosova 9, 191002 St. Petersburg, Russia
2   School of Electrical Engineering, Tel Aviv University, Tel Aviv 69978, Israel; markohen@gmail.com (H.M.); pginzburg@post.tau.ac.il (P.G.)
3   Light-Matter Interaction Centre, Tel Aviv University, Tel Aviv 69978, Israel; roichman@tauex.tau.ac.il
4   School of Chemistry, Tel Aviv University, Tel Aviv 69978, Israel
5   School of Physics & Astronomy, Tel Aviv University, Tel Aviv 69978, Israel
*   Correspondence: i.shishkin@metalab.ifmo.ru

Received: 7 November 2019; Accepted: 9 January 2020; Published: 13 January 2020

Abstract: Advances in laser and optoelectronic technologies have brought the general concept of optomechanical manipulation to the level of standard biophysical tools, paving the way towards controlled experiments and measurements of tiny mechanical forces. Recent developments in direct laser writing (DLW) have enabled the realization of new types of micron-scale optomechanical tools, capable of performing designated functions. Here we further develop the concept of DLW-fabricated optomechanically-driven tools and demonstrate full-3D manipulation capabilities over biological objects. In particular, we resolved the long-standing problem of out-of-plane rotation in a pure liquid, which was demonstrated on a living cell, clamped between a pair of forks, designed for efficient manipulation with holographic optical tweezers. The demonstrated concept paves the way for the realization of flexible tools for performing on-demand functions over biological objects, such as cell tomography and surgery to name just few.

Keywords: holographic optical trapping; direct laser writing

---

1. Introduction

Three-dimensional optical microscopy techniques are invaluable tools in modern biomedical and biophysical studies. Imaging techniques include, for example, confocal [1], multiphoton [2], and super-resolution [3] microscopy. In most cases, these methods require the immobilization of objects under study, at least during image acquisition. A common way to obtain three-dimensional (3D) characterization of the investigated objects is to combine fluorescent dyes and image sectioning. This requires additional sample preparation and, in some cases, may harm or alter the studied system. In biomedical diagnostics, it is especially beneficial to obtain such characterization in a label free manner. For this reason, various 3D tomography techniques, based on quantitative phase microscopy, were developed in the past few years. For example, single-cell optical coherence tomography can be implemented [4], however, imaging-based techniques appear more practical. These techniques require the ability to rotate an object in suspension in a controlled manner. Sample scanning can be achieved by several methods, including loading objects under study in gel-filled microcapillaries and rotating them mechanically [5,6]. Optical tweezers also have been demonstrated as a viable tool for object scanning by using single-beam time-shared trap [7] (suitable only for elongated objects like *E. coli*), and by using multiple optical traps applied to non-spherical objects like diatoms [8] and yeast cells [9].

The possibility of the rotation of an individual cell was demonstrated using a pair of counterpropagating beams from single-mode fibers inserted in the microfluidic channel [10–12], using holographic optical trap [13] or electrorotation [14]. However, it should be noted that trapping of living tissue with counterpropagating laser beams or with structured light in holographic tweezers is constrained by localized heating [15] and phototoxicity [16,17].

Our approach provides capabilities of full 3D manipulation of biological samples within solutions and allows achieving a set of essential functionalities, including (i) prevention of photoinduced damage to living cells, (ii) manipulation of transparent/low contrast objects, (iii) 3D manipulation, including rotation, of spherical species. Furthermore, from the fundamental standpoint, we demonstrate the utilization of radiation pressure forces for achieving the controllable rotation of objects. Our general concept is depicted in Figure 1a, where optomechanically driven 'cell clamps' immobilize a biological cell. Those micron-size clamps are fabricated with the help of direct laser writing [18] (DLW). This technique is based on two-photon absorption [19] in photopolymerizable materials [20], which is an extremely viable method for the fabrication of structures with a sub-micron-scale resolution. A few notable examples of DLW-based structures for opto-fluidic applications include force and topography-sensing optically driven scanning microprobes [21–23], light-actuated microsyringes [24], and platforms for targeted light delivery to microscopic objects like cells [25]. It should be noted that several designs were developed earlier to achieve out-of-plane rotation, like paddlewheel [26] and crankshaft-like structure [27], however none of them have been tested for the manipulation of individual cells.

Our auxiliary structures are driven into motion with the help of holographic optical tweezers. Each clamp is illuminated with three beams—a pair for immobilization and the third one for achieving the rotation of the trapped cell (revolver geometry). Those micro-tools allow for the clamping of an object, translating it towards the analyzing apparatus, rotating it, and finally releasing it back to the suspension. Furthermore, the immobilized cell is not directly illuminated by intense laser light and the whole scheme does not rely on cell's parameters, which makes this approach quite universal. We report on the design, fabrication, and use of 3D printed unique cell clamps that enable trapping, translation, and rotation of cells using optical forces focused away from the cell.

2. Materials and Methods

The proposed approach towards cell rotation is schematically depicted in Figure 1a. A pair of auxiliary tools fabricated by DLW is detached from substrate and immobilized with trapping laser beams. Afterwards the desired cell is located and is immobilized by a pair of tools driven into its proximity simultaneously.

A scanning electron microscopy (SEM) image of the designed cell clamps, shaped like forks, is presented in Figure 1b,c. The structure has several essential elements: (i) a fork end for clamping a cell; (ii) base and top spheres for optical trapping with gradient forces, which are used to control their position and orientation. The radius of the sphere is large enough compared to the fork core to ensure localized trapping [23]. (iii) Three spheres, forming a revolver operating like a windmill, are placed in between these two spheres. The distance between the centers of 'actuating' spheres and the axis of the symmetry of the structure was set 5 μm. The clamps are 30 μm long and each spherical feature is 5 μm in diameter. The distance between the base and top spheres used for immobilization of the tool was chosen to be 18 μm. It should be noted, that the dimensions of the microtool could be reduced roughly by a factor of 2, however a larger size was chosen for better mechanical stability and to allow easier detachment with a micromanipulator. Before performing optical experiments, the coverslips with the microforks were cured overnight with a UV lamp in order to suppress residual fluorescence from the photoinitiator and to increase their mechanical stiffness.

The projection of a defocused trap on the 'revolver' part of the structure is the optimal way to use radiation pressure to rotate the fork around its axis. Rotation can be initiated by turning the third optical trap on, stopped by turning it off, and reversed by projecting the trap of the other side of the fork in real time using a holographic optical tweezers (HOTs) setup [28]. The setup used green 532 nm laser, a reflective spatial light modulator (SLM) module, beam expanders and inverted bright field microscope. The laser beam was expanded in order to overfill SLM aperture. The SLM is placed in the focus of shrinking telescope, which forms 4f-system with the microscope objective. The zero order spot was blocked in the focal plane of the negative beam expander after the SLM. The beam is reflected upward inside the microscope using a beam splitter cube and is focused using a 100× Olympus oil immersion objective (NA = 1.4) into the sample chamber. The phase masks of the SLM for the trapping and manipulating (open, close and rotate) of the micro-tools were designed using MATLAB (Version R2013b, MathWorks, Inc., Natick, MA, USA) and calculated using Gerchberg–Saxton iterative algorithm. The traps placement has been designed to be symmetric with respect to the zero-order beam to reduce the intensity of the higher diffraction orders of the SLM.

Phosphate buffer saline (PBS) with 0.5% TWEEN 20 was used as working medium. The structures were mechanically detached from the coverslip using a glass microneedle connected to the micromanipulator (Scientifica Patchstar, Scientifica, East Sussex, UK) before conducting the experiments. After detaching two structures from the surface, the shutter of the laser was opened. By using the motorized stage the structures are brought into the trapping focal spots of the laser and are immobilized. Since the access for the glass capillary was needed, the samples were not sealed in double-glass chamber and the occasional addition of water was needed in order to compensate for the evaporation.

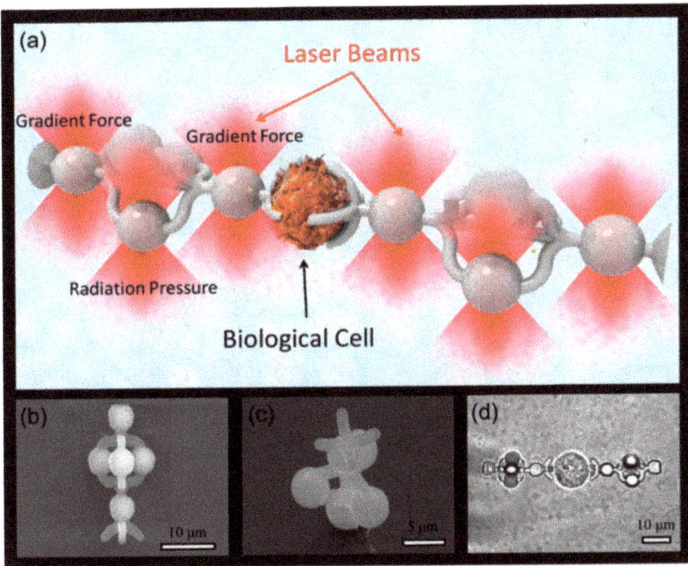

Figure 1. (**a**) An artist's view of the cell clamps at work. The fork-like shaped clamps are optically trapped in 3D by holographic optical tweezers. Scanning electron microscopy (SEM) side view (**b**) and top view (**c**) of fabricated auxiliary microtools. (**d**) Microscope image of microtools engaged in immobilization of sw480 adenocarcinoma cell.

3. Results

In order to demonstrate feasibility of the proposed approach towards object manipulation, it was necessary to show the capability of rotation of the individual micro-tool. After detachment from the coverslip, the tool was immobilized using two generated traps, positioned at anchoring points marked with red crosses in the first frame in Figure 2. After successful immobilization, the third 'actuator' trap was generated the microns off-plane in the spot marked with the cross. The stable trapping of individual tool was achieved with laser power of 0.8 W incident on SLM, which was distributed between three trapping spots. The relative power of the 'actuator' trap was reduced compared to the two main immobilization traps by 50% to improve stability of trapping.

The video sequence of the experiment with a single micro-tool is presented in Figure 2 as set of individual frames captured with 1.5 s interval (see Supplementary Video S1 for complete sequence). It can be clearly seen that, with the proposed configuration of the traps, the desired axial rotation can be achieved. The rotational motion was induced with the radiation pressure force that pushes one of the beads in the revolver part. The photon momentum is transformed to the structure owing to light absorption in the polymer, which arises from residual molecules of photoinitiator and intrinsic material absorption. It is worth noting that translational motion in the plane of view of the trapping objective can be demonstrated straightforwardly, and is not shown here.

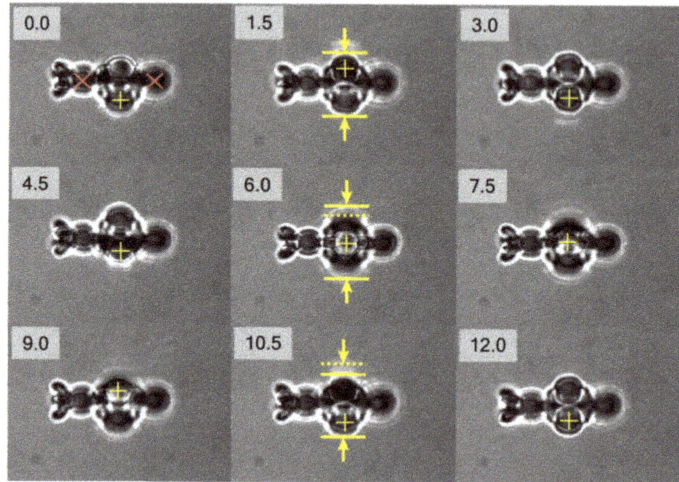

Figure 2. Frame sequence obtained from captured video demonstrating rotation of a trapped single micro-tool. Frames were extracted every 1.5 s. Red crosses in frame at 0 s mark immobilization trap positions. Yellow crosses mark positions of displaced microspheres. The solid and dotted yellow lines shown in frames for 1.5, 6, and 10.5 s reveal variance of one of the dimensions of projection of the microtool captured by camera.

We analyze the rotation dynamics and trapping stability of individual micro-tool by image analysis algorithms. Sufficient contrast between the microtool and the background allows to implement edge detection algorithm. Each captured frame of the recorded video is processed as follows: the edges of the object are detected using a Sobel operator followed by the dilation and filling of gaps in resulting image, which allows to obtain binary mask corresponding to the object. The properties of the resulting binary images were analyzed using Matlab regionprops function. The center-of-mass (CoM) positions were extracted and axes of the equivalent ellipse with same normalized second moments as the original object binary image were obtained. These parameters were further used for analysis of motion of a single microtool.

The information on CoM position over each frame is presented as probability distribution in Figure 3a. The respective data on X and Y position distributions is presented in Figure 3b. The stiffness of trapping potential was analyzed using equipartition theorem and allowed to obtain values of $k_x = 2.22$ pN/µm and $k_y = 1.96$ pN/µm. The values of the trapping potential stiffness could be used for assessment of the maximum possible force which could be exerted on the immobilized cell in order to assess individual cell stiffness [29].

The equivalent ellipse minor axis variance over time is presented in Figure 3c. The periodic variations of the value can clearly be observed, which can be attributed to rotation of the microtool. The corresponding Fourier spectrum of the extracted signal is presented in Figure 3d. Fourier analysis of the time-varying parameter allowed to reveal processes with different periodicity—$T_1 = 8.6$ s

(0.117 Hz), which corresponds to complete 360 degrees revolution of the microtool and $T_2 = 3.3$ s (0.3 Hz) which could be attributed to out-of-plane rotation of the tool by 120°.

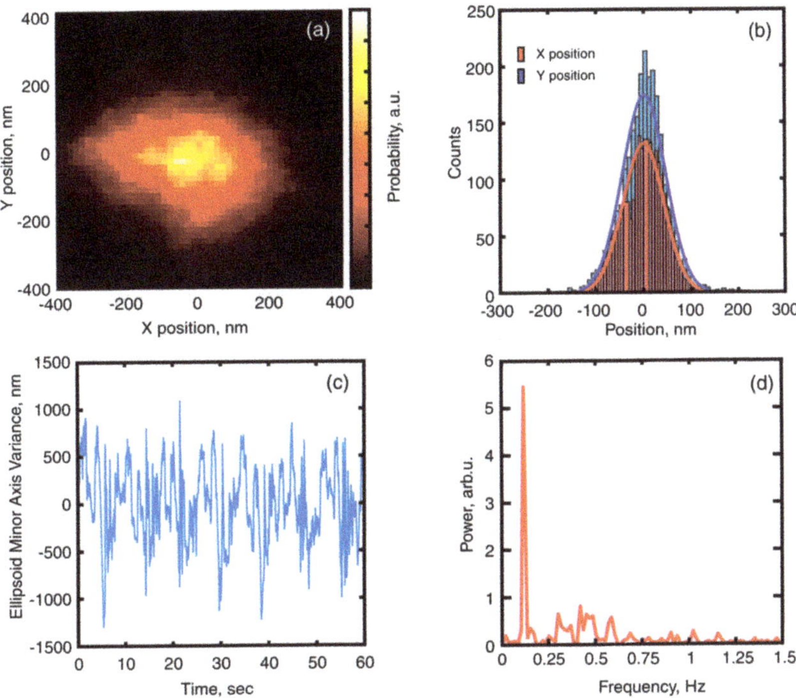

Figure 3. (**a**) Probability distribution of a center of mass (CoM) for single microtool. (**b**) Position distribution for X and Y coordinates of CoM. (**c**) Variance of equivalent ellipsoid minor axis, as a function of time (**d**) Power spectrum of the ellipsoid minor axis variance.

In order to demonstrate proof-of-concept, i.e., the capability of axial rotation of the living biological object, we have undertaken the experiments using yeast cells as the test object. For such an experiment, a pair of micro-tools were detached and immobilized in optical traps with relative separation of 15 microns between edges. After finding the object for studies, the tools were driven together to proximity (video is presented in Supplementary Materials Video S2), resulting in immobilization of the tested object. The success of the immobilization was checked by scanning with the microscope stage. The 'actuator' traps were turned on, resulting in simultaneous rotation of the trapped micro-tools, which transferred the torque on the object clamped between them. The frames of the captured video sequence are presented in Figure 4 (cell immobilization is shown in Supplementary Materials Video S3, cell rotation is shown in Supplementary Materials Video S4).

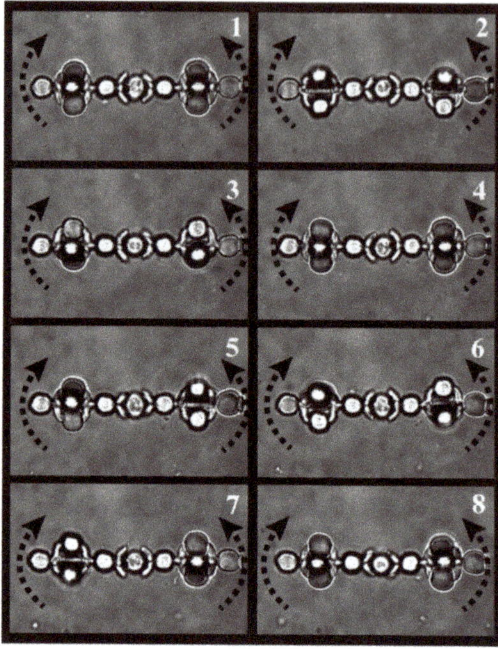

Figure 4. Frame sequence of rotation of trapped yeast cell. Frames were captured each second.

4. Discussion

The recorded videos and their frame-by-frame sequences demonstrate the feasibility of the proposed approach of using auxiliary structures for the micro-manipulation of objects. The analysis of motion of single microtool revealed the stability of microtool immobilization and out-of-plane rotation with the speed of 6–7 revolutions per minute was shown. The proof-of-principle manipulation of an individual live cell with a pair of microtools was demonstrated as well. However, several problems should be addressed in order to successfully implement these microtools for more complex studies.

First, for optical tomography applications, the rotation angle of the sample should be known. This can be achieved by the synchronization of rotation of the auxiliary tools. For this, one needs to project holograms with the out-of-plane position of 'actuator' trap in order to control the angle of revolution of the individual microtool. The current implementation of the trapping algorithm (e.g., static projection of 'actuator' trap) does not provide sufficient control over the rotation speed of the individual tool and does not synchronize the motion of a pair of tools.

Second, the Brownian motion of tools in liquid results in drifts of the studied object not only in the XY-plane, but in the Z-plane as well. The simple image processing technique implemented in this work allowed to determine the centroid position of a single microtool and can be further extended to the analysis of the motion of a pair of microtools with the cell immobilized in between them.

5. Conclusions

In conclusion, we presented a new approach that will allow a complete 360-degree scan of the biological object in-vitro embedded in its host fluid environment. For example, optical diffraction tomography (ODT) [30] allows for the measurement of the refractive index distribution of optically transparent objects, such as cancer cells (our proof of concept result appears in Figure 1d). The method does not require labeling or high intensity light sources. Crucially, the resolution of ODT depends on the range of the angle from which imaging takes place. Two-axis full rotation is optimal. Two principal approaches exist for the scan acquisition in ODT, namely illumination scanning [31] and sample scanning [32]. It should be noted that illumination scanning methods are constrained by limited projection angles [33]. Our proposed technique provides the possibility to performed such two-axis rotation by re-trapping a cell after one-axis rotation is performed. Experiments combining our technique and ODT are underway. These developments can open new horizons in microscopy, where accurate and full three-dimensional mapping of biological objects, and even other valuable functions can be performed with auxiliary optomechanically driven micro-tools.

Supplementary Materials: The following are available online at http://www.mdpi.com/2072-666X/11/1/90/s1, Video S1: 'Single fork rotation'. Video S2: 'Forks clamping'. Video S3: 'Forks cell trapping'. Video S4: 'Yeast rotation'.

Author Contributions: Conceptualization, P.G. and Y.R.; methodology, I.S., H.M. and Y.R.; software H.M. and Y.R.; formal analysis, I.S.; investigation, H.M.; resources, P.G. and Y.R.; data curation, H.M.; writing—original draft preparation, I.S.; writing—review and editing, P.G. and Y.R.; visualization, I.S.; supervision, P.G. and Y.R.; project administration, P.G.; funding acquisition, P.G. All authors have read and agreed to the published version of the manuscript.

Funding: This research was funded by ERC StG 'In Motion'.

Conflicts of Interest: The authors declare no conflict of interest.

References

1. Pawley, J.B. *Handbook Of Biological Confocal Microscopy*; Springer: Boston, MA, USA, 2006.
2. Denk, W.; Strickler, J.H.; Webb, W.W. Two-photon laser scanning fluorescence microscopy. *Science* **1990**, *248*, 73–76. [CrossRef]
3. Egner, A.; Hell, S.W. Fluorescence microscopy with super-resolved optical sections. *Trends Cell Biol.* **2005**, *15*, 207–215. [CrossRef]
4. Choi, W.J.; Park, K.S.; Eom, T.J.; Oh, M.-K.; Lee, B.H. Tomographic imaging of a suspending single live cell using optical tweezer-combined full-field optical coherence tomography. *Opt. Lett.* **2012**, *37*, 2784. [CrossRef] [PubMed]
5. Fauver, M.; Seibel, E.; Rahn, J.R.; Meyer, M.; Patten, F.; Neumann, T.; Nelson, A. Three-dimensional imaging of single isolated cell nuclei using optical projection tomography. *Opt. Express* **2005**, *13*, 4210–4223. [CrossRef] [PubMed]
6. Kus, A.; Dudek, M.; Kemper, B.; Kujawinska, M.; Vollmer, A. Tomographic phase microscopy of living three-dimensional cell cultures. *J. Biomed. Opt.* **2014**, *19*, 46009. [CrossRef] [PubMed]
7. Carmon, G.; Feingold, M. Rotation of single bacterial cells relative to the optical axis using optical tweezers. *Opt. Lett.* **2011**, *36*, 40. [CrossRef]
8. Tanaka, Y.; Wakida, S. Controlled 3D rotation of biological cells using optical multiple-force clamps. *Biomed. Opt. Express* **2014**, *5*, 2341. [CrossRef]
9. Habaza, M.; Gilboa, B.; Roichman, Y.; Shaked, N.T. Tomographic phase microscopy with 180° rotation of live cells in suspension by holographic optical tweezers. *Opt. Lett.* **2015**, *40*, 1881–1884. [CrossRef]
10. Kreysing, M.K.; Kießling, T.; Fritsch, A.; Dietrich, C.; Guck, J.R.; Käs, J.A. The optical cell rotator. *Opt. Express* **2008**, *16*, 16984. [CrossRef]
11. Kreysing, M.; Ott, D.; Schmidberger, M.J.; Otto, O.; Schürmann, M.; Martín-Badosa, E.; Whyte, G.; Guck, J. Dynamic operation of optical fibres beyond the single-mode regime facilitates the orientation of biological cells. *Nat. Commun.* **2014**, *5*, 5481. [CrossRef]
12. Müller, P.; Schürmann, M.; Chan, C.J.; Guck, J. Single-Cell Diffraction Tomography with Optofluidic Rotation about a Tilted Axis. In Proceedings of the SPIE 9548, Optical Trapping and Optical Micromanipulation XII, San Diego, CA, USA, 25 August 2015; p. 95480U.
13. Cao, B.; Kelbauskas, L.; Chan, S.; Shetty, R.M.; Smith, D.; Meldrum, D.R. Rotation of single live mammalian cells using dynamic holographic optical tweezers. *Opt. Lasers Eng.* **2017**, *92*, 70–75.
14. Phys, J.A. A microfluidic chip for single-cell 3D rotation enabling self-adaptive spatial localization A micro fl uidic chip for single-cell 3D rotation enabling self-adaptive spatial localization. *J. Appl. Phys.* **2019**, *126*, 234702.

15. Liu, Y.; Cheng, D.K.; Sonek, G.J.; Berns, M.W.; Chapman, C.F.; Tromberg, B.J. Evidence for localized cell heating induced by infrared optical tweezers. *Biophys. J.* **1995**, *68*, 2137–2144. [CrossRef]

16. Mohanty, S.K.; Sharma, M.; Gupta, P.K. Generation of ROS in cells on exposure to CW and pulsed near-infrared laser tweezers. *Photochem. Photobiol. Sci.* **2006**, *5*, 134–139. [CrossRef] [PubMed]

17. Konig, K.; Liang, H.; Berns, M.W.; Tromberg, B.J. Cell damage in near-infrared multimode optical traps as a result of multiphoton absorption. *Opt. Lett.* **1996**, *21*, 1090–1092. [CrossRef] [PubMed]

18. Kawata, S.; Sun, H.B.; Tanaka, T.; Takada, K. Finer features for functional microdevices. *Nature* **2001**, *412*, 697–698. [CrossRef] [PubMed]

19. Goppert-Mayer, M. Uber Elementarakte mit zwei Quantensprungen. *Ann. Phys.* **1931**, *401*, 273–294. [CrossRef]

20. Maruo, S.; Nakamura, O.; Kawata, S. Three-dimensional microfabrication with two-photon-absorbed photopolymerization. *Opt. Lett.* **1997**, *22*, 132–134. [CrossRef]

21. Phillips, D.B.; Simpson, S.H.; Grieve, J.A.; Bowman, R.; Gibson, G.M.; Padgett, M.J.; Rarity, J.G.; Hanna, S.; Miles, M.J.; Carberry, D.M. Force sensing with a shaped dielectric micro-tool. *EPL* **2012**, *99*, 58004. [CrossRef]

22. Gibson, G.M.; Bowman, R.W.; Linnenberger, A.; Dienerowitz, M.; Phillips, D.B. A compact holographic optical tweezers instrument. *Rev. Sci. Instrum.* **2012**, *8*, 113107. [CrossRef]

23. Phillips, D.B.; Padgett, M.J.; Hanna, S.; Ho, Y.L.D.; Carberry, D.M.; Miles, M.J.; Simpson, S.H. Shape-induced force fields in optical trapping. *Nat. Photonics* **2014**, *8*, 400–405. [CrossRef]

24. Villangca, M.J.; Palima, D.; Bañas, A.R.; Glückstad, J. Light-driven micro-tool equipped with a syringe function. *Nat. Publ. Gr.* **2016**, *5*, e16148-7. [CrossRef] [PubMed]

25. Palima, D.; Bañas, A.R.; Vizsnyiczai, G.; Kelemen, L.; Ormos, P.; Glückstad, J. Wave-guided optical waveguides. *Opt. Express* **2012**, *20*, 2004–2014. [CrossRef] [PubMed]

26. Asavei, T.; Nieminen, T.A.; Loke, V.L.Y.; Stilgoe, A.B.; Bowman, R.; Preece, D.; Padgett, M.J.; Heckenberg, N.R.; Rubinsztein-Dunlop, H. Optically trapped and driven paddle-wheel. *New J. Phys.* **2013**, *15*, 63016. [CrossRef]

27. Phillips, D.B.; Gibson, G.M.; Bowman, R.; Padgett, M.; Rarity, J.G.; Carberry, D.M.; Hanna, S.; Miles, M.J.; Simpson, S.H. Fashioning Microscopic Tools. In Proceedings of the Optics in the Life Sciences, Washington, DC, USA, 14–18 April 2013; p. TM2D.3.

28. Grier, D.G.; Roichman, Y. Holographic optical trapping. *Appl. Opt.* **2006**, *45*, 880–887. [CrossRef]

29. Guck, J.; Ananthakrishnan, R.; Mahmood, H.; Moon, T.J.; Cunningham, C.C.; Käs, J. The optical stretcher: A novel laser tool to micromanipulate cells. *Biophys. J.* **2001**, *81*, 767–784. [CrossRef]

30. Sung, Y.; Choi, W.; Fang-Yen, C.; Badizadegan, K.; Dasari, R.R.; Feld, M.S. Optical diffraction tomography for high resolution live cell imaging. *Opt. Express* **2009**, *17*, 266. [CrossRef]

31. Choi, W.; Fang-yen, C.; Badizadegan, K.; Oh, S.; Lue, N.; Dasari, R.R.; Feld, M.S. Tomographic phase microscopy. *Nat. Methods* **2007**, *4*, 717–719. [CrossRef]

32. Charrière, F.; Marian, A.; Montfort, F.; Kuehn, J.; Colomb, T.; Cuche, E.; Marquet, P.; Depeursinge, C. Cell refractive index tomography by digital holographic microscopy. *Opt. Lett.* **2006**, *31*, 178. [CrossRef]

33. Kawata, S.; Nakamura, O.; Minami, S. Optical microscope tomography I Support constraint. *J. Opt. Soc. Am. A* **1987**, *4*, 292. [CrossRef]

 *micromachines*

Review

Optical Micromachines for Biological Studies

Philippa-Kate Andrew [1], Martin A. K. Williams [2,3] and Ebubekir Avci [1,3,*

[1]   Department of Mechanical and Electrical Engineering, Massey University,
   Palmerston North 4410, New Zealand; k.andrew@massey.ac.nz
[2]   School of Fundamental Sciences, Massey University, Palmerston North 4410, New Zealand;
   m.williams@massey.ac.nz
[3]   MacDiarmid Institute for Advanced Materials and Nanotechnology, Wellington 6140, New Zealand
*   Correspondence: e.avci@massey.ac.nz

Received: 21 January 2020; Accepted: 9 February 2020; Published: 13 February 2020

Abstract: Optical tweezers have been used for biological studies since shortly after their inception. However, over the years research has suggested that the intense laser light used to create optical traps may damage the specimens being studied. This review aims to provide a brief overview of optical tweezers and the possible mechanisms for damage, and more importantly examines the role of optical micromachines as tools for biological studies. This review covers the achievements to date in the field of optical micromachines: improvements in the ability to produce micromachines, including multi-body microrobots; and design considerations for both optical microrobots and the optical trapping set-up used for controlling them are all discussed. The review focuses especially on the role of micromachines in biological research, and explores some of the potential that the technology has in this area.

Keywords: optical tweezers; multi-component micromanipulators; radiation damage; life sciences; optical microrobots

---

1. Introduction

Improvements in tools for the visualisation of objects at the micro- and nano- scale have given researchers the ability to investigate materials and processes previously out of reach. The dominant forces and phenomena that can be accessed at these small scales often differ greatly from those observed at the macroscopic scale. This can lead to challenges, such as adhesion between objects in micro-manipulation experiments [1,2], but also to opportunities, where forces which would be unnoticeable at the macroscale can be used to great effect on micro and -nanoscale objects; such as in the atomic force microscope and optical and magnetic tweezers [3]. Of these technologies, optical tweezers have shown themselves to be particularly flexible tools for the investigation of structures, systems and processes covering the mesoscale. This has led to optical tweezers being used as a tool in the vast fields of biomedical research, material science and soft-matter studies, where they can be used to apply and measure forces that range from 10^{-14} to 10^{-10} Newtons. Optical tweezers were invented by Nobel laureate Arthur Ashkin and colleagues and were first officially introduced in a 1986 paper [4] following more than a decade of experiments with what was termed the radiant pressure force [5–10]. These experiments were themselves inspired by earlier work by researchers in the early-to-mid 1900s [11–14], who had experimentally shown what James Clerk Maxwell had theorised: that light carries momentum. However, these earlier researchers had been severely limited by the range of forces that could be applied, and these experiments relied on observation of extremely slight perturbations. This restriction was removed with the invention of the laser [15–17], which allowed for experiments with high intensity beams of coherent light. The potential for applications other than simple manipulation of dielectric spheres was soon realised, with a paper by Ashkin and

Dziedzic demonstrating the ability to manipulate viruses and bacteria shortly after the influential paper that introduced single-beam optical traps [18]. This led to rapid uptake of the technology by researchers interested in probing the properties of biological subjects. In particular, the relationship between force and biological behaviours has been a subject of enduring research interest, as researchers took note of the ability to investigate intra-cellular phenomena non-invasively [19]. A particularly interesting subset of optical tweezers research is the emerging field of optical microrobotics. Optical microrobotics has been demonstrated to have applications in optical scanning-force microscopy [20], cell manipulation [21–23] and cell analysis [24]. The behaviour of optical micromachines has also been used to validate the use of classical mechanical models at the microscale [25,26] as well as providing insight into how material properties may be affected by scale [27]. However, optical micromachines have been under-utilised thus far in biological studies, making this area one with tremendous potential for researchers. In order to place the opportunities for micromachines in context, this review covers a brief introduction to optical tweezers and the impact of optical trapping in biological studies, as well as exploring the concerns associated with optical trapping of biological specimens. The potential for optical micromachines in this area is then explored through an evaluation of the achievements and challenges associated with optical micromachines, considerations for microrobot design and the choice of optical tweezers for manipulation of optical microrobots. Finally, the review finishes with a speculative view of the future, briefly exploring a few possibilities for the use of optical microrobots in biological studies.

2. An Introduction to the Theory of Optical Tweezers

Although the theory of optical tweezers is not a central feature of this review, an introduction to the optical force is helpful. Many papers have been written about the theory involved in modelling optical tweezers [28–30], and this summary does not hope to offer more than a basic introduction, with the reader encouraged to follow the references for more thorough explanations. While the effect of the sun on direction of comet tails prompted Johannes Kepler's theories about the mechanical effects of light, the real impact and potential of radiation pressure are most clearly observed on the micro- and -nano-scale, where the intensity afforded by lasers makes the forces originating from changes in the momentum of light significant. The main feature of an optical tweezers setup is a laser focussed through a high numerical aperture (NA) objective, which turns the force associated with the laser into a useful tool. The high NA objective serves to tightly focus the laser beam to a narrow region of space, resulting in an extremely steep gradient of the laser's electromagnetic field at this point. This is shown in Figure 1. The force associated with the focussed laser beam is treated in two parts in the theory of optial tweezers: a gradient force that "pulls" objects in towards the focus, and a scattering force which acts like water in a "hosepipe" of photons, bombarding the object [31].

Calculations of the gradient and scattering forces can then be performed according to different equations depending on whether the object being trapped can be said to belong to the ray optics regime (particle diameter d » trapping wavelength λ) or the Rayleigh regime (d « λ). The ray optics explanation assumes that the objects are sufficiently large that they act as lenses, and relies on the understanding that the collimated laser beam is comprised of many single rays of light. The intense focus of the laser beams used for optical trapping means that there is a higher density of rays at the focus. Each ray carries momentum and experiences a change in this momentum when refracted through the microsphere. This change in momentum corresponds to an equal and opposite change in the momentum of the trapped object, acting to move the object. The moment provided by the beam is calculated by decomposing the incident light beam into a series of rays with appropriate intensity, direction and polarisation, and each ray is understood to be a plane wave which changes direction upon interaction with the object's surface. Diffraction is ignored in this regime, and the gradient (F_{grad}) and scattering (F_{scat}) force can be calculated using Equations (1) and (2), respectively [16,32].

$$F_{grad} = \frac{n_1 P}{c} R sin(2\theta) - \frac{T^2[sin(2\theta - 2r) + R sin(2\theta)]}{1 + R^2 + 2R cos(2r)} \tag{1}$$

$$F_{scat} = \frac{n_1 P}{c} 1 + R cos(2\theta) - \frac{T^2[cos(2\theta - 2r) + R cos(2\theta)]}{1 + R^2 + 2R cos(2r)}. \tag{2}$$

In these equations the incident momentum is given by $\frac{(n_1 P)}{c}$ where n_1 is the refractive index of the trapping medium, P is the power of the ray, and c is the speed of light. The angle of incidence is given by θ and the angle of refraction by r. The terms R and T refer to the Fresnel coefficients for reflection and transmission due to the interface of the medium and target object, which are dependent on the polarisation of the beam [33]. For the case where the light is unpolarised, equal amounts transverse-magnetic (P polarized, R_p, T_p) and transverse-electric (S polarized, R_s, T_s) are used. The Fresnel coefficients for transverse-magnetic and transverse-electric waves are given by Equations (3)–(6).

$$R_p = |\frac{(n_1\sqrt{1 - (\frac{n_1}{n_2}sin(\theta_i))^2} - n_2 cos(\theta_i))}{(n_1\sqrt{1 - (\frac{n_1}{n_2}sin(\theta_i))^2} + n_2 cos(\theta_i))}|^2 \tag{3}$$

$$R_s = |\frac{(n_1 cos(\theta_i) - n_2\sqrt{1 - (\frac{n_1}{n_2}sin(\theta_i))^2}}{(n_1 cos(\theta_i) + n_2\sqrt{1 - (\frac{n_1}{n_2}sin(\theta_i))^2}}|^2 \tag{4}$$

$$T_p = 1 - R_p \tag{5}$$

$$T_s = 1 - R_s. \tag{6}$$

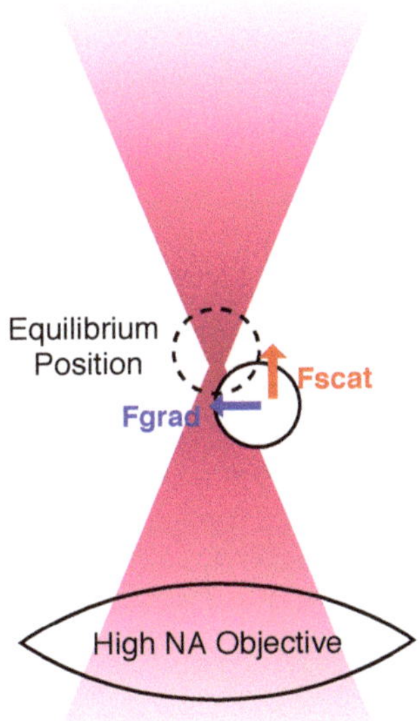

Figure 1. Optical tweezer theory treats the force associated with the focussed laser beam as two components, which can be visualised as resulting from the intense focussing of the light.

If the ray optics regime can be thought of as treating objects as lenses, then objects meeting the size requirements to be considered as Rayleigh scatterers can be thought of as particles that become polarised in response to the changing electromagnetic field. The interaction of the induced dipoles with the steep gradient of the electromagnetic field associated with the tightly focussed laser results in the object being pulled towards the waist of the beam [34]. The scattering force can then be understood as being produced because the intensely focussed field is not static, and thus it induces a magnetic field around the dipoles, resulting in the particles being "pushed" [29].

In this regime the equations for the gradient and the scattering forces are given by (7) and (8) [35,36], where n_{med} is the refractive index of the medium, c is the speed of light, a is the radius of the object being trapped, m is the ratio of the refractive indices of the object and medium ($m = \frac{n_{obj}}{n_{med}}$), I is the intensity, r is the object's position (x, y, z) and k is the wave number ($k = \frac{2\pi}{\lambda}$). The intensity of the laser is related to the electromagnetic field E by Equation (9). These equations provide reasonable approximations of the forces acting on spheres ranging from a few nanometres to a few hundred nanometres, but become unreliable as the size of the object of interest approaches the wavelength of the trapping laser [35].

$$F_{grad}(\boldsymbol{r}, t) = \frac{2\pi n_{med} a^3}{c}\left(\frac{m^2 - 1}{m^2 + 2}\right)\nabla I(\boldsymbol{r}, t) \tag{7}$$

$$F_{scat}(\boldsymbol{r}, t) = \frac{8\pi n_{med} k^4 a^6}{c}\left(\frac{m^2 - 1}{m^2 + 2}\right) I(\boldsymbol{r}, t) \tag{8}$$

$$I(\boldsymbol{r}, t) = \frac{\epsilon_0 c E(\boldsymbol{r}, t)^2}{2}. \tag{9}$$

Unfortunately, most of the objects that researchers are interested in trapping have characteristic dimensions that lie between these two basic regimes, in the intermediate range close to the trapping wavelength. Theoretically, for a small, spherical object in a vacuum, Maxwell's stress tensor provides a solution for the force imparted by an electromagnetic wave, but it assumes complete knowledge of the field, and is impractical for experimental applications. In the case of spherical objects, Mie theory can be used for an exact solution to the problem of scattering [37,38], but it assumes a plane wave, and this makes it inapplicable to the focused beams used for optical tweezers. However, early experiments with laser light validated the use of Mie theory for objects in the intermediate range, with Mie scattering patterns analysed for non-homogeneous, sphere-within-a-sphere particles [39]. This provided valuable information about how this type of particle interacted with the laser light, and provided some evidence that this theory would hold for experiments with biological cells, which cannot always be assumed to be spherical, isotropic or homogeneous structures. Modern use of Lorenz-Mie theory has seen it adapted to cope with arbitrary illumination [40,41], and in this form it is known as Generalised Lorenz-Mie theory (GLMT). GLMT has been combined with other theories of light propagation to more accurately model particular set-ups; such as the Debye-Wolf theory in order to model the forces from holographic optical tweezers [42]. The main limitation of GLMT is that it is applicable to homogeneous, isotropic spheres, rather than complex shapes. Therefore, further generalisation of the theory to extend it to arbitrarily shaped particles was necessary, resulting in the development of T-matrix theory.

First proposed by Waterman in 1965 [43], the T matrix method is useful for calculating the forces on an object in a dynamic optical tweezers set-up, as the matrix depends on the shape, size, refractive index and position of the scattering object, rather than the incident or outgoing fields of light. Additionally, the matrix simplifies for objects with some degree of symmetry, with a sphere being the simplest shape. This greatly reduces the computational effort. The T-matrix method has been reviewed and utilised by many researchers since Waterman's introduction of the method for solving scattering problems, with different methods for calculating the matrix used [44], including the extended boundary condition method (EBCM) [45] and the point-matching method [46,47]. A comprehensive overview of the T matrix method specifically for optical tweezers has been written by Nieminen et al. [48] and a thorough explanation of T-matrix and generalised Lorenz-Mie theory has also been produced by Gouesbet [49].

While these methods provide a way to calculate the forces associated with optical tweezers, researchers can reduce difficulty by modelling optical tweezers as a Hookean spring [50]. The Hookean spring model is based on the understanding of an optical trap made from a Gaussian beam as being a "harmonic potential well", with a defined equilibrium point, and a characteristic stiffness constant. Using the separation of the scattering and gradient components of the force as a starting point, in a stable trapping scenario the gradient force will dominate, meaning that the particle will remain at an equilibrium position in a trap, notwithstanding Brownian motion. Then, the stiffness coefficient of the optical trap can be estimated by measuring the root mean square displacement of the particle from the centre of the trap induced by thermal fluctuations and using the equipartition theorem [51,52]. Other calibration methods for measurement of trap stiffness have been developed [53], and are popularly based on balancing a viscous drag imposed on the particle in the trap [54] or by performing power spectrum analysis [55]. The simplicity of the Hookean spring approximation, and the wealth of different methods for calibration hints at a limitation of this method: the effective spring constant is affected by fluctuations in experimental parameters such as viscosity and temperature, and so may require recalibration during long-running experiments. For researchers looking to bypass this issue, a method of force measurement arising from first principles has been successfully demonstrated in both dual-beam, counter-propagating optical traps [56] and single beam traps [57]. This method involves the measurement of the change in momentum of light before and after interaction with the sample, and requires the collection of scattered light in order to extract the change in momentum from the resulting intensity pattern using a position-sensitive detector. A drawback of this method is that it requires precise alignment of all the components required, and it is not feasible to collect all of the scattered light in a single-beam set-up. However, the results from Reference [57] indicate that the "lost" backward-scattered light contributes very little to the overall calculations, and this could be a useful method for researchers working with non-spherical or non-homogeneous objects, as it does not require experiment-specific calibration.

3. First Applications and Impact on Fields

While experiments with the radiant pressure force were initially intended for atomic cooling applications [9,31], their potential for use in biological research was quickly realised [18,58,59]. Ashkin's own experiments with trapping viruses and bacteria [18] sparked further interest in optical tweezers as a non-invasive technique for probing the material properties of biological subjects [60] and for investigating force-related phenomena [61]. In the past few decades this research has included applications that have ranged from investigating the mechanisms behind cell stiffening due to high blood pressure [62] to examining the impact of chemical binding on DNA's mechanical properties [63–65]. In fact, stretching of single-molecules such as DNA has proven to be a topic that has benefited greatly from the development of optical tweezers.

While the pioneering DNA stretching experiment was conducted using magnetic tweezers [66], the spatial resolution of optical tweezers, which, even in the 1990s, allowed displacements of only a few nanometres to be affected [67], proved particularly attractive to researchers for single molecule studies. Additionally, the range of forces that can be applied cover a range from 10^{-14}–10^{-10} N, which means that several different force regimes can be investigated [68], from those where thermal fluctuations provide restoring forces, to those where high-forces that can impart conformational changes in constituents of the chain [54]. The popularity of DNA stretching studies has meant that there is ample data available for researchers wanting to develop improved models for the molecule's structure and behaviour [69]. In this way, theoretical models for single-molecule structure and behaviour can be readily evaluated, and their limitations found. An example of this is the breakdown of the freely-jointed model for single DNA molecules and the subsequent validation of the worm-like chain model [69–72].

Early optical tweezers-based DNA stretches featured molecules that were fixed to a coverslip at one end, with a free-moving end attached to a functionalised microbead. This microbead was then held in the optical trap and the coverslip was translated in order to create a stretch on the molecule [73]. However, this meant that the stretch was affected by disturbances to the translating equipment, due to mechanical coupling of the stage to the rest of the set-up. Additionally, tethering the DNA to the coverslip meant that the stretches could not be truly uni-directional, with an orthogonal component that could not be determined with a level of accuracy comparable to the in-plane displacement. Another popular method for fixing one end of the DNA tether was attaching the corresponding functionalised microbead to a micropipette, which likewise introduces the possibility of unwanted mechanical coupling. A potential fix for the problem of mechanical coupling can be found in the introduction of steerable traps. Steerable traps allow for the manipulation of both ends of a free-floating DNA molecule, decoupling the process from the rest of the apparatus. In the resulting "dumbbell assay", functionalised beads are attached to DNA molecules that have been modified to feature bead-binding compounds (typically biotin and digoxigenin) at the ends. Each bead is then confined to a separate optical trap, with one trap being stepped away from the other, which is held in a constant position, in order to stretch the molecule [74]. The change in the distance between the beads is taken as the extension of the DNA tether, allowing for the use of the Hookean spring model of optical tweezers. These three different methods for forming a DNA tether for stretching can be seen in Figure 2, which has been reproduced from Heller et al.'s review of the use of optical tweezers for analysing DNA-protein complexes [75].

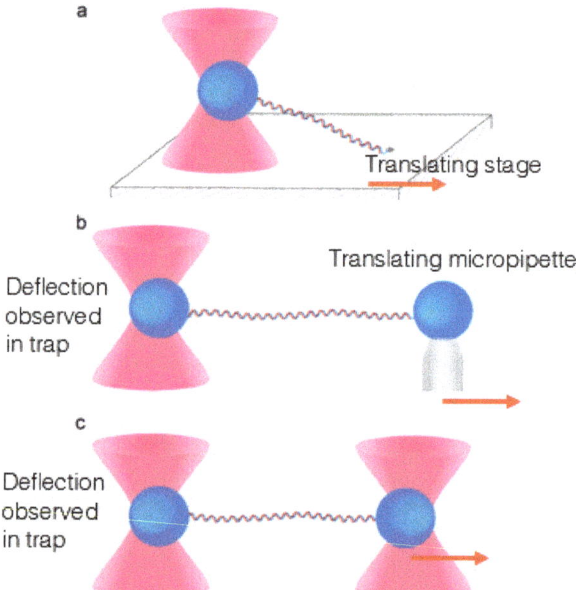

Figure 2. Three different methods for stretching a DNA tether with optical tweezers can be seen in this figure: the translating slide method (**a**), using an anchoring micropipette (**b**) and using two separate optical traps (**c**). Reproduced with permission from Reference [75]. ©2014 American Chemical Society.

While the introduction of a second optical trap in DNA stretching studies decoupled the experiment from vibrations of the apparatus, it also introduced another source of intense laser light. It is accepted that laser irradiation has a negative impact on trapped biological samples, but it is not known how much the laser light affects the DNA molecule, as it is not being directly trapped. However, this leads to an important topic; concerns for biological research involving optical tweezers.

4. Optical Trapping: Concerns for Biological Research

The negative effects of laser radiation on biological samples were discovered just as quickly as the potential for optical tweezers in biological studies [18], necessitating the development of strategies for minimising damage to biological samples during assays with optical tweezers. The first consideration for making the trapping beam less damaging to samples was the beam wavelength, with the beam changed from visible radiation to infrared in order to reduce absorption [18]. Infrared lasers commonly produce light in the near-infrared range from 700 nm to around 1200 nm. Even within this range there are significant differences in the damage observed in biological samples, depending on the wavelength used [76,77]. Interestingly, while 740–760 nm have been found to be the most damaging, the commonly used 1064 nm wavelength was found to reduce clonability of CHO cells to below 20% of the control group within five minutes of laser exposure [77]. 990 nm, on the other hand, was found by the same study to only reduce clonability to around 70% after a full 20 minutes of exposure, at the same laser power. However, when the damage caused to cellular DNA after exposure for 30s is compared, 1064 nm is clearly less damaging than wavelengths in the range of 700–900 nm [76]. Therefore, the wavelength used for trapping can be clearly seen to be one of the key considerations when trapping biological samples.

In another study, the presence of oxygen was considered as a factor influencing damage, with the motility of E. coli examined to quantify damage [78]. A range of wavelengths from 830 nm to 1064 nm were used, and the wavelength-dependent damage observed appeared to agree closely with that from Liang et al. [77]. When free oxygen was removed from the system, either through introduction of a scavenging species, or by growing and trapping cells in an anaerobic environment, the damage was found to decrease to levels comparable with the control group, which was not exposed to the laser. Other studies have been performed that also seem to corroborate the hypothesis that oxygenation of the sample is a main cause of damage induced during optical trapping [79,80]. However, there is not yet a clear consensus on the exact mechanism for this damage, and other effects related to laser proximity have also been found to induce damage, such as localised heating [81,82]. Research into parameters influencing the damage inflicted by optical trapping has revealed that the trapping medium affects the temperature increase the laser generates, with glycerol producing much higher temperature variation than water [82].

Not all experiments with biological matter can be adapted to remove oxygen, change the trapping medium, or cope with the introduction of a scavenging species. Therefore, other methods for reducing damage as much as possible must be considered. The laser beam itself has been identified as a potential hazard to biological samples, simply due to the high intensity needed to produce an optical trap. As the electromagnetic field varies sharply, with the intensity dropping rapidly with respect to the distance from the waist of the beam, it has been proposed that some of this damage can be avoided simply by spatially separating the optical traps from the object being examined. An example of this theory can be found in the use of functionalised microbeads for DNA stretching, where the DNA itself is not trapped [69] but rather it is chemically bonded to microbeads which act as trapping points for the molecule. The use of these beads is also necessary to enable DNA stretching at all, as the size of the molecule is below the diffraction limit; a limiting factor for optical tweezers until recently [83]. This also has additional advantages, one of which is that the size of the beads can be adjusted to a size that is advantageous for optical trapping with the set-up being used, ensuring a high trap stiffness. The process can also be better visualised with the help of these beads, with the behaviour of the microbeads informing the researcher about the behaviour of the molecule being stretched between them. If direct trapping of the DNA molecule was to be attempted, then the use of a dye would be required, in order to determine the location of the molecule. Use of such dyes can change the mechanical properties of the DNA molecule, depending on the binding mechanism [61], and labelled molecules can be affected by photobleaching, severely limiting experiment times [84]. While using quantum dots as fluorescent markers may solve the issue of photobleaching, there are concerns about toxicity, although this is not considered a drawback for in vitro studies [85].

Larger biological samples such as cells and their organelles can be directly trapped without difficulty [20], but it is desirable to find strategies for mitigating the damage that high intensity laser traps can inflict [86]. As in the case of single-molecule studies, the effect of the wavelength of light used has been considered, with near-infrared traps preferred because of low endogenous absorption at wavelengths between 700 and 1200 nm [87]. Additionally, the use of non-Gaussian traps has been explored as a possible method for stable trapping of living cells. This includes the creation of "optical funnels" using ring-shaped optical fibres to create a funnel of light, in which objects with a lower refractive index than the trapping medium can be stably trapped [88]. Annular traps have also been constructed using axicons, and used to trap living sperm cells, with the added benefit of exposing the spermatozoa to less intense light [89]. Annular traps constructed using axicons have the added benefit of propagating for long distances within the sample, unlike traps constructed using Gaussian beams which are limited to a small distance from the objective. This means that the working distance in the z-direction can be much greater in a ring-shaped trap [90]. The use of two lenses to create the Bessel beams required also means that the size of the resulting ring-shaped beam can be adjusted by moving one of the lenses, changing the focal length and thus the ring's diameter, which has potential for studies involving objects of different sizes [91]. However, while some experimental success has been demonstrated using axicons for trapping cells and atoms [92], the use of axicon traps is less straightforward than the theory suggests [93], as the lower intensity of the light consequently means that the optical forces produced are lower than in Gaussian-beam traps. This limits the applications of these traps, as they cannot produce the high forces required for applications such as investigating the mechanical properties of cells [94]. This means that alternative solutions that are not reliant on changing the power or intensity of the trapping beam are desirable for applications that require high forces.

A possible method for retaining the effective power of optical traps, while reducing damage to the subject is using other optically trapped objects as intermediates, effectively using them as end-effectors to manipulate delicate biological subjects. The use of functionalised beads for DNA stretching can be regarded as pioneering work in indirect optical trapping, with functionalised rigid-body structures made from SU-8 also used to manipulate cells in a system with six degrees of freedom [21]. Indirect optical manipulation of cells has also been demonstrated using non-functionalised end-effectors [22], indicating that non-specific bonding due to physical adhesion forces can be sufficient for such studies. This indicates that intermediate optically trapped objects may have potential for a variety of applications where orientation of the cell is important, such as cell aspiration and nucleus extraction. Additionally, experiments using trapped objects as "force probes" indicate that optical end-effectors can be designed to retain the sensitivity of force application that optical tweezers are known for [95,96]. This is important, as the range and resolution of force that can be applied by optical tweezers are distinguishing features of the technology when compared to conventional mechanical manipulation, where limited resolution of force and displacement can limit the use of these methods for high-precision tasks [97].

5. Optical Microrobots: Potential Tools for Biological Studies

While researchers have not yet reached a consensus on the exact mechanisms for damage incurred during optical trapping, it is widely agreed that exposure to intense light is a problem. As previously mentioned, it is not always possible to adapt experiments for lower trapping powers, and it is not always practical for researchers to have multiple custom optical trapping set-ups for different samples. Therefore, using microrobots to reduce specimen contact with the trap's focus, while still enabling application of optical trapping forces, would be a conceptually simple solution for the problem of optically inflicted damage. As already noted, the use of optically trapped objects as scanning probes, for DNA stretching, and as tools for simple cell rotation can be thought of as pioneering work in the field of optical microrobotics. It follows that the use of more complex, articulated robots may be the solution to reducing exposure to laser light, and potentially they may also be the answer to overcoming the limitations of optical tweezers, such as limited force and difficulty performing truly 3D manipulation.

While the potential for optical microrobotics in biological research is clear, there are still relatively few examples of the technology used in this field. This could be partially attributed to historical difficulties in manufacturing optical microrobots. Micro-manufacturing methods have traditionally been iterative processes based off those used for semi-conductor applications, limiting the resolution and shapes that can be achieved. While there is some scope for the use of these iterative-process methods in the development of microrobots [98], they are limited due to their two-dimensional nature. In the case of the paper by Mittas et al., the difficulty of using traditional silicon micromanufacturing techniques can be seen due to the need of precise alignment of the different patterning stages. Additionally, the use of etching for high aspect-ratio features is problematic as it is a time dependent process, which means that the etchant generally spreads in an undesirable manner, leading to feature-undercut. However, this undercutting has been purposefully used to produce three-dimensional micromachine elements, when etchant has been used to remove patterned elements from silicon wafers to produce independent components for use with optical tweezers [99]. In situations where repeatable manufacturing and high feature resolution (<1 μm) are required, relying on etching to produce the desired outcome becomes impractical. Similarly, photoresist patterning using mask aligners can be used to good effect to produce reasonably complex structures [100], but this is an arduous task that is prone to error due to the need for extremely precise alignment of subsequent layers, and only allows for 2.5 dimensions, with extrusion of a planar pattern occurring in one direction. Additionally, making multi-component robots from these methods is complex, due to the requirement for assembly, although the components themselves can be produced [101]. Fortunately, the development of three-dimensional photolithography, based on two-photon absorption polymerisation, has enabled the manufacture of complex structures with <100 nm resolution. The use of this laser-based 3D printing technology is well-established, with the first patent for 3D stereolithography granted in 1986 [102], and examples of complex structures present in literature since the early 2000s [103].

Although two-photon absorption was first theoretically suggested by Nobel laureate Maria Göppert-Mayer in her doctoral thesis [104], like many light-based phenomena it was only experimentally demonstrated after the invention of the laser. Two-photon absorption involves the absorption of two photons- that can be of the same or different energies- to bridge the energy gap between electronic states of a molecule and bring it into a higher energy electronic state. The especially remarkable feature of two-photon absorption is the speed at which the second photon must be absorbed to prevent the molecule in question returning to its initial state from the intermediate, higher-energy virtual state it occupies following the absorption of the first photon. The fact that two photons must be absorbed for the process to occur means that it is non-linear, with the rate of absorption (R) being given by (10), where δ is the absorption cross-section and I is the intensity of the light.

$$R = \delta I^2. \tag{10}$$

This phenomenon is the cause of two-photon absorption polymerisation (TPAP), which occurs when two-photon absorption occurs in a matrix of monomers or oligomers and a photoacid-generator [105]. This initiates polymerisation of the material in the regions of the material where the required energy threshold is met [106]. A highly focussed, femto-second pulsed laser beam is used to provide the necessary light for the reaction, and due to the narrow region of focus, high aspect ratio features with feature resolution of less than 100 nm is possible, as the region of polymerisation can be finely controlled [107,108]. This has enabled the printing of complex structures with fine detail, with functional 3D-printed helical springs being evidence of what the technology can achieve [24,109]. This technology is especially promising for creating optical microrobots for use in biological studies, as several commercial resists are biocompatible, with SU-8 being a common example. Additionally, conceptually any resin can be used, so long as it has sufficient absorption of the wavelength of the laser used to induce TPAP, and a high enough proportion of photoacid in the matrix. This allows researchers to create custom resins with the properties that they require, a feature that has already been exploited by researchers creating cell-scaffolds from hydrogels [110]. While in principle two-photon absorption polymerisation simply requires a high-speed pulsed laser, the invention of the Nanoscribe, a commercial solution developed by researchers at Karlsruhe Institute of Technology, has led to more user-friendly TPAP-based 3D printing. The availability of commercial solutions such as this means that 3D laser printing has become accessible to researchers from different backgrounds, widening the scope of applications.

6. Optical Microrobots: Achievements to Date and Challenges

It has been established that complex shapes can be printed using TPAP, and there has been some work performed using optically trapped objects as end-effectors for studies involving the manipulation of biological subjects [111]. This positions optically driven micromachines as a promising technology for researchers looking to reduce exposure to intense laser traps. However, there are also several challenges that need to be addressed before the technology can be adopted for regular use, rather than as a novelty.

Optically-driven microrobots have been successfully used to amplify forces from optical tweezers [24,25], and to transform planar input-motion to out-of-plane rotation [112]. It has also been demonstrated that trapped microspheres, arranged in groups of three, create an attractive pocket that approximates a weak optical funnel; which could be used to indirectly trap biological objects [113]. This is similar to what can be achieved using the "doughnut" shaped beams used for dark trapping [88], and demonstrates the potential for optical microrobots as an alternative and more flexible solution to altering the beam used for optical trapping. Additionally, the impressive resolution and dimensions that can be achieved using TPAP has enabled the printing of nano-springs. These nano-springs have been used as more than just examples of what TPAP can achieve, they have also been used to measure force amplification [24], and to investigate the impacts of post-curing UV exposure on photoresins [114], using optical tweezers. The use of levers and articulated microrobots to amplify force and produce out-of-plane movement shows that micromachines can be used to improve upon the abilities of optical tweezers, which are currently rather limited in these areas. Additionally, by demonstrating that classical models of springs hold at the microscale [24,26], researchers have shown that springs could potentially be used to directly measure the forces applied with optical tweezers. However, this approach requires extensive knowledge of the polymer used to fabricate the springs and precise control of printing parameters, as non-uniformity could lead to large differences from expected results [115]. This has become less of a problem as laser-based 3D printing technology has advanced, and as more research has been performed to ascertain the effects that laser printing parameters have on mechanical properties of printed objects [116], but it is still worth considering.

While these examples show what can be achieved in optical microrobotics, the authors also make note of the challenges involved in producing and actuating them. Particular difficulties are caused by adhesion forces between discrete moving parts [25,112] and the tolerance of pin joints when optical trapping is performed. Adhesion forces in particular form a barrier when it comes to developing articulated optical microrobots, as more complex machines offer more opportunities for seizure due to links and joints bonding to each other through physical interactions such as van der Waals forces. Additionally, this poses a particular barrier for the use of microrobots in biological research, where studies are often carried out in particular fluids which are intended to mimic in vivo conditions. Some of these fluids, such as the tris-buffered saline used for DNA stretching [117], have relatively high salt concentrations, leading to a shortening of the Debye length and thus a decreasing importance of any stabilising electrostatic interactions [118]. Methods for reducing adhesion forces in microrobotics are based around reducing contact area between objects [112,119]. The reason for this can be easily appreciated when one considers the equations for the van der Waals force between a sphere and a plane (11) [120]—a key contributor to adhesion forces.

$$F_{vdW} = \frac{H}{6}\frac{d}{2z} + \frac{d}{2(z+d)^2} - \frac{1}{z} + ln(\frac{1}{z+d}).\tag{11}$$

In Equation (11), H represents the Hamaker constant of the material [121], d is the diameter of the microsphere and z is the separation between the sphere and the plane. This equation assumes the same material for the plane and the microsphere, and that the plane is extremely large compared to the microsphere, and so has some limitations, but it clearly illustrates the effects of separation distance and size of the objects on the van der Waals forces between them. Therefore, it can be deduced that adhesion forces can be limited by increasing separation between moving parts of the nanomachines, and by minimising the contact area between parts, which is supported by results from the literature.

Another challenge in creating multi-component optical microrobotics is being able to use non-spherical objects. Relatively small ($d \approx \lambda$) spherical shapes are the easiest objects to trap with single Gaussian traps, as they are isotropic and the curved surface leads to lower reflection and so to lower scattering forces when compared to flatter objects. Additionally, as already mentioned, they are also the easiest shape to model. Conversely, objects with high aspect ratios, such as cylinders, tend to orient themselves with the axis of the optical trap [122], a tendency first discovered when Ashkin and Dziedzic trapped cylindrical viruses [17], which can lead to difficulty in manipulating such objects as desired. A method for getting around this problem is to add spherical features as trapping "handles" [123], similar to the use of functionalised beads in DNA stretching. This strategy has been adopted by many researchers, and spherical parts feature in many optical microrobot designs [23,25,112,124]. Another strategy is to use several optical traps, positioned so that the forces are balanced when manipulation takes place, regardless of the shape of the "handles". Examples of this can be found in the use of cylindrical trapping points in the development of an optical scanning probe [125], where spheres are used as tracking points rather than trapping handles, and in the use of flat discs as handles for structures produced using 2D deposition and etching methods [99].

While the gradient force is most commonly used to control the movement of optical microrobots, optical tweezers can also actuate micromachines through the scattering force, or through thermal effects. The use of the scattering force to trap objects has perhaps most successfully been used in the case of the counter-propagating optical trap, where two aligned, counter-propagating laser beams work together to create a balanced trap based on the scattering force produced by each laser [126,127]. This configuration, while complex compared to single-beam optical tweezers, allows for lower intensity light to be used, and so does not require the high NA objectives used for a gradient trap. The use of a lower NA objective also allows for longer working distances than optical tweezers, which are generally limited to a working distance of less than a millimetre from the objective. On the other hand, an example of the use of thermal effects from optical tweezers to actuate microtools is the creation of a "micro-syringe", where the inclusion of a thin metal layer inside a 3D printed chamber creates thermally-induced flow in response to the laser light, moving silica and polystyrene microspheres in the process [128]. The process of pulling in a microsphere to later be ejected can be seen in Figure 3.

As discussed previously, the force resulting from the interaction of the laser beam with the object of interest is commonly split into two parts, and $F_{grad} > F_{scat}$ is commonly accepted to be the condition necessary for stable trapping. However, the scattering force can also be used to actuate microrobots, and can be used for applications involving a rotating part which can be "pushed" by the light, rather than being dragged into place [129,130]. In-plane rotation of a twin-rotor micropump has been demonstrated, by focusing a Gaussian beam on the sides of flat-lobed "wings" [131]. Out of plane rotation has also been demonstrated using an optical "paddle wheel". In this case two micro-spherical features were used as "trapping handles" to hold the object in place, while the central "paddle" was driven by the scattering force of another trap, creating out-of-plane rotation [132]. A schematic of this microrobot, as well as stills taken from a video of it moving, can be seen in Figure 4.

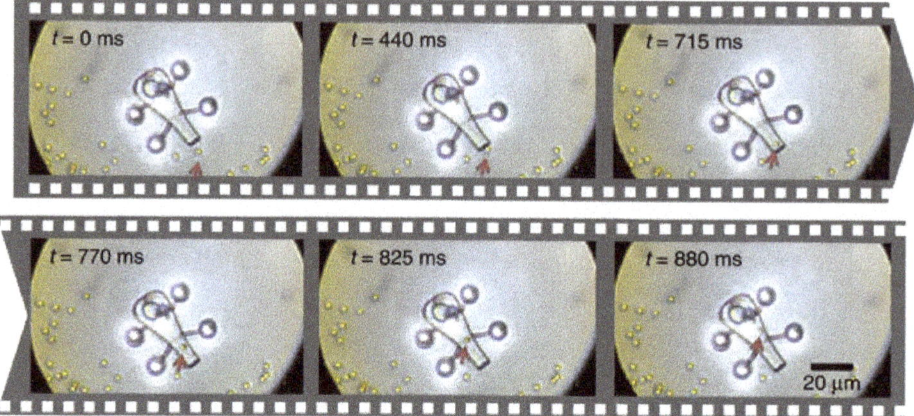

Figure 3. The inclusion of a thin metallic layer inside a 3D printed chamber leads to the movement of silica and polystyrene payload particles through convection. Reproduced from Villangca et al. [128], under Creative Commons license NC ND 4.0.

Optical forces have also been utilised to rotate chiral objects, where the design of the object means that the incoming momentum is unbalanced and creates an optical torque around the beam's trapping axis [133]. This has been utilised by researchers to produce micropumps, through the rotation of self-orienting Archimedes screws [134], and through the use of rotors with tilted "blades", similar to a waterwheel [135]. Anisotropic quartz particles have been used to apply and measure torque from Gaussian, single-beam optical tweezers, due to the birefringence of quartz resulting in non-uniform response to linearly polarised light [136]. Another paper by the same group demonstrated the influence of changing the beam polarisation on the particle's resulting torque, showing that the torque obtained depends on the polarisation rotation rate, allowing for the use of a trapped particle as a passive constant-torque wrench [137]. In the case of optical rotors, it is generally true that the rate of rotation is governed by the power of the laser used, and these could be used for micropump applications such as cell-sorting or as mechanisms for exerting mechanical force on objects such as cells, to investigate mechano-transduction processes.

Control of articulated optical microrobotics is commonly achieved by manually moving them using an optical trap. Therefore, the development of automated control strategies presents a gap in the research, although some examples exist for simple microrobotics set-ups involving rigid bodies [138,139]. Improving the control strategies available for articulated optical micromachines, particularly ones that allow for parallelising and automating, would help to move them from research curiosities to useful tools for non-experts. The ability to do this is strongly linked to the ability to repeatably produce micromachines, as tiny differences in structure can have large effects at the microscale, which could impact functionality. This was noted in Reference [112], where success rates of different designs were tracked in order to determine a reliable microrobot design. The ability to model the behaviour of microrobots in response to the beam is also important for control system design, and repeatable experimental results are integral to building reliable models. However, a major difficulty that researchers face is the fact that all parts of an optical microrobot are affected by the presence of an optical trap. This means that there is difficulty controlling the movement of multi-component microrobots, as the different components are all attracted to a single point, the optical trap. Therefore the shape of the micromachine's components, and the mechanical coupling of the different parts is key to ensuring the micromachine operates as required. Additionally, as covered earlier in this review, the number of optical traps that can be used, and the ability to control them separately should be considered.

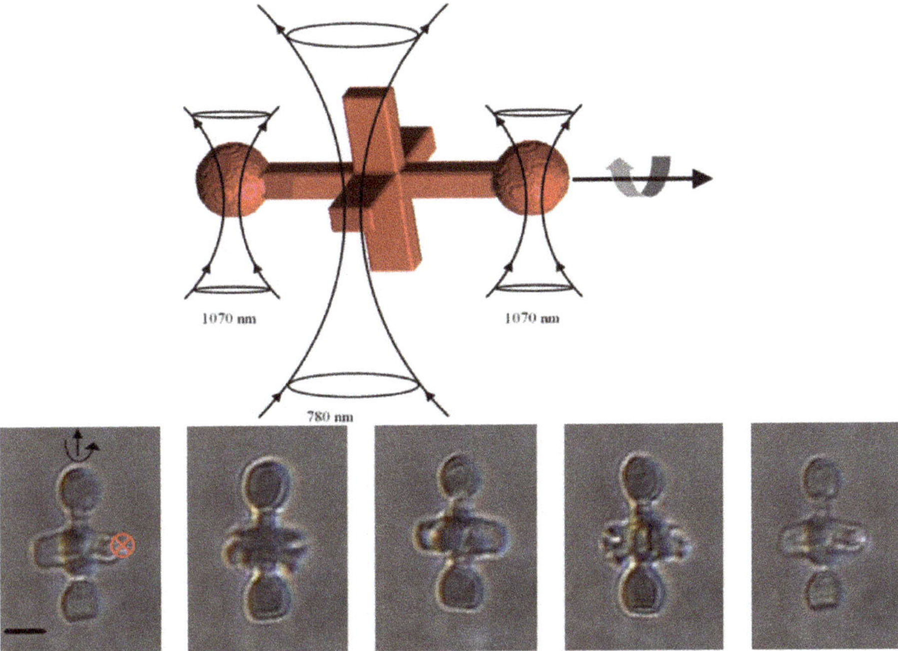

Figure 4. Asavei et al. produced an optical "paddle wheel" held in place using two 1070 nm optical traps and driven with a separate 780 nm beam. Reproduced from "Optically trapped and driven paddle-wheel", [134], under Creative Commons license CC BY 3.0. (**Top**): Schematic of the device. (**Bottom**): Stills taken from a video of the device in action, where the position of the 780 nm beam is shown with a red x.

7. Effects of Microrobot Shape

As previously touched on in this review, the simplest shape for optical trapping is the homogeneous sphere. However, obviously this is not an ideal shape for multi-body microrobots, for tasks involving gripping, or for tasks where movement around a fixed axis is the desired outcome. Therefore, the shape of the microrobot must be decided according to the same principles as any other engineering challenge: namely, the task at hand and equipment restrictions with respect to manufacturing and controlling the resulting machine.

The production of simple, classical machines involving lever arms, springs and even Archimedes screws has been achieved using TPAP, and functionality has been demonstrated by using optical tweezers to actuate movement. The shape of these microrobots is highly relevant when it comes to their functionality. In the case of an optically controlled lever arm [24], the long arm, if unconstrained by the pin-joint, would naturally orient itself with the trapping beam. The fixed axis, which is attached to the substrate, prevents this from occurring to some extent, but the effect has been noted as a restriction of micro-scale pin joints, as it increases undesirable, out-of-plane movement [25]. The inclusion of spherical features which act as "trapping handles" serves to reduce undesirable motion, as these points are more strongly drawn into the optical trap than relatively flat surfaces.

While this review has already covered some of the situations in which the scattering force is useful for actuating a microrobot, the shape of micromachines that are intended to be actuated in this way also deserves some attention. For instance, the angle of interaction of the optical tweezers with the surface of the screw is vitally important for producing rotation in the desired direction. This has been demonstrated in the creation of the wheel with tilted blades referenced earlier [135], where the direction of rotation was changed by moving the optical tweezers to different points of the wheel. This effect was exploited in the creation of a twin-rotor using a similar design by different researchers [140]. This can be simply explained according to the refraction and reflection of light in an asymmetric shape leading to the resultant forces being directed off-centre and leading to an optical torque [133]. This concept also explains the positive impact of adding spherical "trapping handles" to micromachines, as the curved surface means that the refractive (gradient) and reflective (scattering) components of force from the optical tweezers are directed through the centre of the sphere.

While spherical handles work well for optical tweezers utilising Gaussian beams, where the refractive index of the micromachine is higher than that of the surrounding medium, this is not the case for situations where the refractive index of the object is the lower value of the two. In this case, in fact, the opposite becomes true, and doughnut-shaped objects have been stably trapped using Gaussian optical tweezers, utilising the interaction of the gradient force on the inner walls of the doughnut [141,142]. A diagram of this can be seen in Figure 5, where the case of a spherical particle (a) is compared with ring-shaped particles (b) and (c), and the impact of the beam waist position in the Z-axis is evaluated. The demonstration of stable optical trapping of low-refractive index particles further increases the options for optical trapping studies by increasing the range of mediums that can be used, and it is possible that ring-shaped objects could be used as "trap handles" for micromachines in these scenarios.

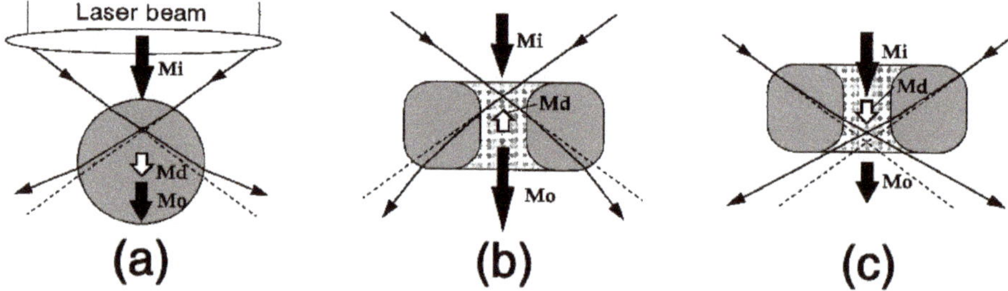

Figure 5. Illustration of the interaction of the optical tweezers with differently shaped low-refractive index particles, and the impact of the position of the beam waist. In (**a**) and (**c**) the object shape and the beam waist position act to push the object out of the trap, whereas in (**b**) the incident forces are balanced on the inner walls of the ring, and the position of the beam waist in the Z direction acts to pull the object towards the waist, producing a stable trap. *Mi* is the incident momentum, *Mo* is the outgoing momentum and *Md* is the momentum transferred from the light to the object. Reproduced with permission from Reference [141] ©The Optical Society.

8. Choice of Optical Tweezers for Microrobotics

It can be seen that the ability to create multiple optical traps is extremely useful in microrobotics. This is particularly true in the case of free-floating micromachines with moving parts [112,132], when the overall position of the microrobot must be controlled as well as its relative motion. Using several different laser beams would be both expensive and greatly increase the space required for an optical trapping set-up. Luckily, several solutions have been developed for the creation of multiple traps from a single focused laser beam. These can be divided into space-sharing and time-sharing set-ups, with holographic optical tweezers [143] being an example of the first type and fast-switching acousto-optic deflectors [144] being an example of the latter. Holographic optical tweezers (HOT) utilise spatial light modulators (SLM) to split an incoming beam into several different beams, which are then directed through the microscope aperture, as shown in Figure 6. This technique is particularly flexible in terms of the number of traps that can be created, and the range of movement that is allowed, as it relies only on the pattern imposed on the SLM. Work regarding the calibration of HOT using the power spectrum analysis has also revealed that trap stiffness does not vary significantly during trap movement, and that trap placement is controllable with single-nanometer resolution [145]. These capabilities identify

HOT as a useful and versatile tool for force application and measurement in biophysical experiments. However, the real-time use of the technique is limited, due to the calculations required to determine the required pattern of orientations of liquid crystalline pixels across the device in order to create the desired trap positions. Several different methods for performing this calculation are presented in the literature, including the high-performance yet low-speed Gerchberg-Saxton algorithm [146] and the low-precision yet high-speed direct superposition method [147]. Therefore, when calculating holograms for optical tweezers or other laser-based applications, a trade-off must be made in terms of either speed or accuracy [148], with attempts made to adapt these existing algorithms to produce high speed, high performance alternatives [149,150].

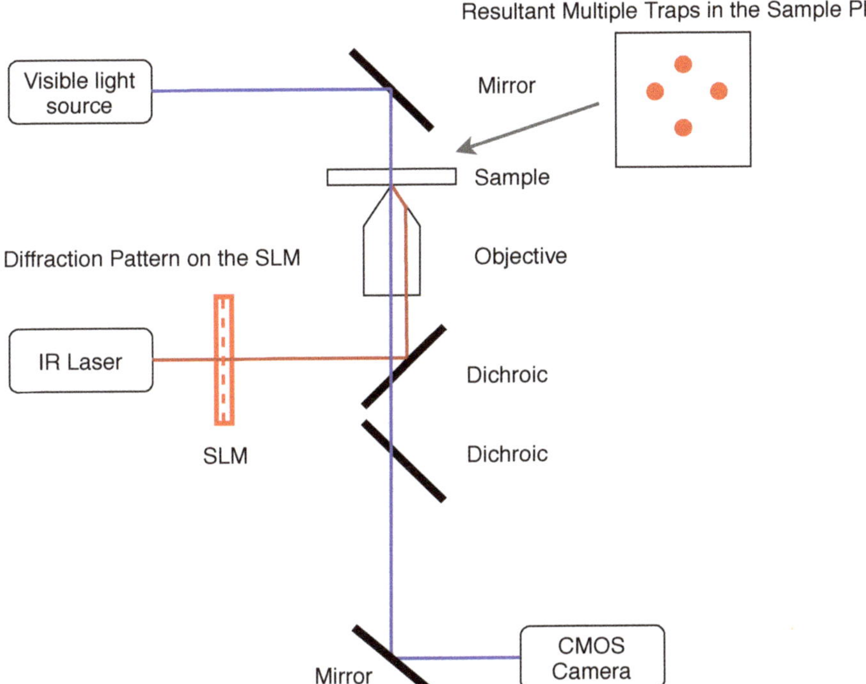

Figure 6. Simple schematic of holographic optical tweezers, based on the authors' laboratory set-up.

Time-sharing set-ups for optical tweezers are based on rapid movement of the beam and typically use acousto-optic (AOD) or electo-optic deflectors (EOD), which deflect the beam by a certain angle, controlled by the frequency of an input signal [151,152]. The calculations involved in this type of dynamic optical tweezers are less time-intensive than those for HOT, but time-sharing optical tweezers are limited in the displacement of traps that can be achieved. This limitation has been tackled through the introduction of additional deflectors [153], and through careful selection of the transducer used to convert the signal that controls the deflector position, which can affect both resolution and range of frequencies that can be used [154]. As HOT and time-sharing dynamic optical tweezers both have their shortcomings, and both have been used successfully in optical microrobotics applications, the decision to use one or the other is largely based on hardware restrictions, with the introduction of an SLM into an existing general optical tweezers set-up potentially much simpler than the creation of a beam-deflecting system. Additionally, commercial options are available for both HOT and AOD-based optical tweezers, but in laboratories where a small range of specific experiments are performed then it may be preferable to build a custom optical trapping set-up according to specific requirements, and researcher expertise. Additionally, many advances have been made in the development of novel optical tweezers setups, and while they are not a central part of this review it is worth mentioning the counter-propagating optical trap [99,127], and the optical fibre tweezers [155]. The latter allows for light to be precisely directed into a sample by use of optical fibres which are pig-tailed to the laser diode, focusing the light using the tapered end of the fibres. Optical fibre tweezers offer the opportunity to precisely trap objects in crowded samples, and so could be considered for situations where trapping is to take place in a complex environment. A review of optical fibre tweezers that covers the use of the technology for a range of optical trapping applications is provided by Zhao and colleagues, including a theoretical explanation of both dual and single fibre setups [156].

9. Looking to the Future

There is potential for the use of micromachines in almost any discipline where optical tweezers are used to perform tasks, rather than as a laser-physics demonstration. Also, as previously mentioned, the improved and simplified ability to produce optical microrobots without post-manufacture assembly means that researchers from a variety of backgrounds can make use of the opportunities that optical microrobots offer. However, the achievements in microrobotics to date appear to have a bent towards biological research, where the dexterity and precision of optical microrobotics could make activities such as polar-body biopsy or cell enucleation swifter and easier. One of the main reasons that microrobotics have such potential in these areas is the potential for parallelising experiments. For instance, cell aspiration is typically performed using a very thin micropipette, which makes it a delicate task that is susceptible to any disruption to the apparatus [157]. Using a micro-machine that features a sharp point, similar to the nanowire used for temperature measurement in Reference [24] to probe cell stiffness could improve experiment success rates by reducing equipment breakage and decoupling the aspiration set-up from outside vibration, and this concept is shown very simply in Figure 7. Using multiple micromachines controlled by multiple optical traps could increase the speed of such experiments, with many aspirations taking place in parallel with one another. Additionally, the highly customisable and controllable nature of optically controlled microrobots in terms of output force and object displacement means that experiments adapted for optical tweezers using optically controlled microrobots could have lower intrinsic uncertainty than conventional, non-optically actuated methods. A potential example could be single-cell surgery, where the exact force required to breach the cell membrane without causing undesirable damage could be carefully applied with optical microrobots every time. As well as this, the potential that optical microrobots have for reducing laser-induced damage during optical tweezer assays further cements their probable application in biological research in the future, as these assays get adapted to the technology. An example of this could be micro-machine assisted DNA stretching, by adapting the protocol that already makes use of functionalised microbeads, as discussed earlier in this review.

The methods developed to control optical microrobotic systems could also indirectly benefit research using optical tweezers. For example, the work done to determine position and orientation in a 3D workspace [158], or to control the position of holographic optical traps in real-time [139,159] has applications in optical tweezer-based cell sorting. Therefore, it is fair to say that research into optical microrobotics will offer both direct and indirect benefits to the scientific community, moving forward.

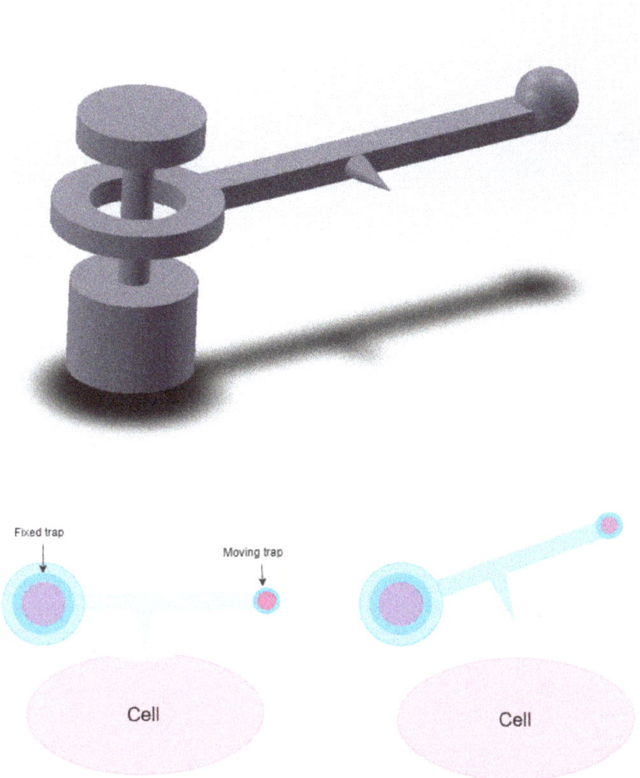

Figure 7. A free-floating microrobot, such as this lever, equipped with a sharp tip, could be used to apply amplified forces for measuring cell membrane stiffness. At least two traps are necessary for this, one fixed trap to hold the microlever in place near the cell and another to move the lever arm.

10. Conclusions

The use of optical tweezers in biological studies has allowed for the investigation of processes and subjects which would otherwise have remained out of reach, due to the resolution of force and displacement possible. However, as has been outlined in this review, the negative effects of optical tweezers are well-known. The common methods for reducing optical damage, such as altering trapping wavelength or introducing radical scavenging species do not allow for flexibility in the experiments that can be performed, and may require completely changing the trapping set-up. This means that such methods are unsuitable for retrofitting to existing trapping set-ups, and are not suitable for general trapping set-ups which may be used for a variety of applications where wavelengths that are more damaging to biological samples are completely acceptable. Reducing the power of the laser is also often less than ideal, as this lessens the forces that can be applied, which limits the ability to research high-force phenomena such as DNA overstretches. Optical micromachines present a flexible solution to these problems, as they can be used to limit direct exposure of the subject to the laser, which has been theorised to be one of the main ways to reduce optically induced damage, through direct sample heating, photobleaching and generation of free radicals in close vicinity to the sample. While indirect manipulation through microrobotics will not remove the risk of damage due to free-radical damage, it would move the origin further away from sensitive biological matter, such as DNA, resulting in more time before their diffusion to the polymer. Additionally, optical micromachines show potential to be used to extend the capabilities in biological studies, with proof of concept demonstrated for force amplification and direct force measurement. The resolution that can be achieved through two-photon absorption polymerisation, along with the ability to print articulated machines also means that many different micromachine designs are possible. This not only means that the technology is extremely flexible and can be adapted for many different applications, but it also means that there could one day be solutions for many of the challenges associated with the technology, such as the dominance of adhesion forces. However, there is still much work to be done in making micromachines attractive to non-roboticists, such as in automating manipulation and improving repeatability. Overall, optical micromachines present many opportunities for biological studies, as well as interesting engineering challenges, and will likely be a subject of research interest for years to come.

Author Contributions: E.A. and M.A.K.W. supervised the project and acquired funding. P.-K.A., wrote the original draft. P.-K.A., E.A. and M.A.K.W. edited the final paper. All authors have read and agreed to the published version of the manuscript.

Funding: Supported by the Marsden Fund Council from Government funding, managed by Royal Society Te Aparangi (MAU1714).

Acknowledgments: The authors thank Urs Staufer and Daniel Fan for their mentoring during P-K Andrew's research stay at the Technische Universiteit Delft.

Conflicts of Interest: The authors declare no conflict of interest.

References

1.	Avci, E. Dynamic releasing of biological cells at high speed using parallel mechanism to control adhesion forces. In Proceedings of the 2014 IEEE International Conference on Robotics and Automation (ICRA), Hong Kong, 31 May–5 June 2014; IEEE: Hong Kong, China, 2019; pp. 3789–3794.
2.	Gauthier, M.; Alvo, S.; Dejeu, J.; Tamadazte, B.; Rougeot, P.; Régnier, S. Analysis and Specificities of Adhesive Forces Between Microscale and Nanoscale. *IEEE Trans. Autom. Sci. Eng.* **2013**, *10*, 562–570. [CrossRef]
3.	Neuman, K.C.; Nagy, A. Single Molecule Force Spectroscopy: Optical tweezers, magnetic tweezers and atomic force microscopy. *Nat. Methods* **2014**, *10*, 491–505. [CrossRef] [PubMed]
4.	Ashkin, A.; Dziedzic, J.M.; Bjorkholm, J.E.; Chu, S. Observation of a single-beam gradient force optical trap for dielectric particles. *Opt. Lett.* **1986**, *11*, 288–290. [CrossRef] [PubMed]
5.	Ashkin, A. Acceleration and Trapping of Particles by Radiation Pressure. *Phys. Rev. Lett.* **1970**, *24*, 156–159. [CrossRef]
6.	Ashkin, A.; Dziedzic, J.M. Optical Levitation by Radiation Pressure. *Appl. Phys. Lett.* **1971**, *19*, 283–285.,1653919. [CrossRef]
7.	Gordon, J.P.; Ashkin, A. Motion of atoms in a radiation trap. *Phys. Rev. A* **1980**, *21*, 1606–1617. [CrossRef]
8.	Ashkin, A. Applications of Laser Radiation Pressure. *Science* **1980**, *210*, 1081–1088. 210.4474.1081. [CrossRef]
9.	Ashkin, A. Trapping of Atoms by Resonance Radiation Pressure. *Phys. Rev. Lett.* **1978**, *40*, 729–732. [CrossRef]
10.	Ashkin, A.; Dziedzic, J.M. Optical Levitation of Liquid Drops by Radiation Pressure. *Science* **1975**, *187*, 1073–1075. [CrossRef]
11.	Nichols, E.F.; Hull, G.F. A Preliminary Communication on the Pressure of Heat and Light Radiation. *Phys. Rev.* **1901**, *13*, 307–320. [CrossRef]
12.	Nichols, E.F.; Hull, G.F. The Pressure Due to Radiation (Second Paper). *Phys. Rev.* **1903**, *17*, 26–50. [CrossRef]
13.	Jones, R.V.; Richards, J.C.S.; Lindemann, F.A. The Pressure of Radiation in a Refractive Medium. *Proc. R. Soc. Lond.* **1954**, *221*, 480–498. [CrossRef]
14.	Lebedev, P.N. Experimental Examination of Light Pressure. *Ann. Phys.* **1901**, *301*, 433–458.
15.	Schawlow, A.L.; Townes, C.H. Infrared and Optical Masers. *Phys. Rev.* **1958**, *112*, 1940–1949. [CrossRef]
16.	Kogelnik, H.; Li, T. Laser Beams and Resonators. *Appl. Opt.* **1966**, *5*, 1550–1567. [CrossRef] [PubMed]
17.	Draegert, D. Single-diode end-pumped Nd:YAG laser. *IEEE J. Quantum Electron.* **1973**, *9*, 1146–1149. [CrossRef]

18. Ashkin, A.; Dziedzic, J.M. Optical Trapping and Manipulation of Viruses and Bacteria. *Science* **1987**, *235*, 1517–1520. [CrossRef]

19. Ashkin, A. Force generation or organelle transport measured in vivo by inrared laser trap. *Nature* **1990**, *348*, 346–348. [CrossRef]

20. Ikin, L.; Carberry, D.M.; Gibson, G.M.; Padgett, M.J.; Miles, M.J. Assembly and force measurement with SPM-like probes in holographic optical tweezers. *New J. Phys.* **2009**, *11*, 023012:1–023012:8. [CrossRef]

21. Xie, M.; Chen, S.; Mills, J.K.; Wang, Y.; Liu, Y.; Sun, D. Cell out-of-plane rotation control using a cell surgery robotic system equipped with optical tweezers manipulators. In Proceedings of the 2016 IEEE International Conference on Information and Automation (ICIA), Ningbo, China, 31 July–4 August 2016; IEEE: Ningbo, China, 2016; pp. 103–108.

22. Aekbote, B.L.; Fekete, T.; Jacak, J.; Vizsnyiczai, G.; Ormos, P.; Kelemen, L. Surface-modified complex SU-8 microstructures for indirect optical manipulation of single cells. *Biomed. Opt. Express* **2016**, *7*, 45–56. [CrossRef]

23. Hu, S.; Hu, R.; Dong, X.; Wei, T.; Chen, S.; Sun, D. Translational and rotational manipulation of filamentous cells using optically driven microrobots. *Opt. Express* **2019**, *27*, 16475–16482. [CrossRef] [PubMed]

24. Fukada, S.; Maruyama, H.; Masuda, T.; Arai, F. 3D fabrication and manipulation of hybrid nanorobots by laser for single cell analysis. In Proceedings of the 2012 International Symposium on Micro-NanoMechatronics and Human Science (MHS), Nagoya, Japan, 4–7 November 2012; pp. 479–481.

25. Lin, C.-L.; Lee, Y.-H.; Lin, C.-T.; Liu, Y.-J.; Hwang, J.-L.; Chung, T.-L.; Baldeck, P.L. Multiplying optical tweezers force using a micro-lever. *Opt. Express* **2011**, *19*, 20604–20609. [CrossRef] [PubMed]

26. Jeong, Y.J.; Lim, T.W.; Son, Y.; Yang, D.-Y.; Kong, H.-J.; Lee, K.-S. Proportional enlargement of movement by using an optically driven multi-link system with an elastic joint. *Opt. Express* **2010**, *18*, 13745–13753. [CrossRef]

27. Ushiba, S.; Masui, K.; Taguchi, N.; Hamano, T.; Kawata, S.; Shoji, S. Size dependent nanomechanics of coil spring shaped polymer nanowires. *Sci. Rep.* **2015**, *5*, 17152:1–17152:8. [CrossRef] [PubMed]

28. Bui, A.A.M.; Stilgoe, A.B.; Lenton, I.C.D.; Gibson, L.J.; Kashchuk, A.V.; Zhang, S.; Rubinsztein-Dunlop, H.; Nieminen, T.A. Theory and practice of simulation of optical tweezers. *J. Quant. Spectrosc. Radiat.* **2017**, *195*, 66–75. [CrossRef]

29. Nieminen, T.A.; Stilgoe, A.B.; Heckenberg, N.R.; Rubinsztein-Dunlop, H. Approximate and exact modelling of optical trapping. In Proceedings of the SPIE NanoScience + Engineering, 2010, San Diego, CA, USA, 27 August 2010; pp. 1–8.

30. Nieminen, T.A.; Heckenberg, N.R.; Rubinsztein-Dunlop, H. Computational modelling of optical tweezers. In Proceedings of the Volume 5514, Optical Trapping and Optical Micromanipulation, Optical Science and Technology, the SPIE 49th Annual Meeting, Denver, CO, USA, 2–6 August 2004; pp. 514–523.

31. Ashkin, A. History of optical trapping and manipulation of small-neutral particles, atoms and molecules. *IEEE J. Sel. Top. Quantum Electron.* **2000**, *6*, 841–856. [CrossRef]

32. Ashkin, A. Forces of a single-beam gradient laser trap on a dielectric sphere in the ray optics regime. *Biophys. J.* **1992**, *61*, 569–582. [CrossRef]

33. Lvovsky, A.I. Fresnel Equations. In *Encyclopedia of Optical Engineering*; Hoffman, C., Driggers, R., Eds.; Taylor and Francis: New York, NY, USA, 2013; pp. 1–6.

34. Neuman, K.C.; Block, S.M. Optical Trapping. *Rev. Sci. Instrum.* **2004**, *75*, 2787–2809. [CrossRef]

35. Harada, Y.; Asakura, T. Radiation forces on a dielectric sphere in the Rayleigh scattering regime. *Opt. Commun.* **1996**, *124*, 529–541. [CrossRef]

36. Smith, P.W.; Ashkin, A.; Tomlinson, W.J. Four-wave mixing in an artificial Kerr medium. *Opt. Lett.* **1981**, *6*, 284–286. [CrossRef]

37. Mie, G. Beiträge zur Optik trüber Medien, speziell kolloidaler Metallösungen. *Ann. Phys.* **1908**, *330*, 377–445. [CrossRef]

38. Salandrino, A.; Fardad, S.; Christodoulides, D.N. Generalized Mie theory of optical forces. *J. Opt. Soc. Am. B* **2012**, *29*, 855–866. [CrossRef]

39. Ashkin, A.; Dziedzic, J.M. Observation of light scattering from nonspherical particles using optical levitation. *Appl. Opt.* **1980**, *19*, 660–668. [CrossRef]

40. Xu, F.; Ren, K.; Gouesbet, G.; Gréhan, G.; Cai, X. Generalized Lorenz-Mie theory for an arbitrarily oriented, located and shaped beam scattered by a homogeneous spheroid. *J. Opt. Soc. Am. A* **2007**, *24*, 119–131. [CrossRef] [PubMed]

41. Ren, K.; Gréhan, G.; Gouesbet, G. Prediction of reverse radiation pressure by generalized Lorenz-Mie theory. *Appl. Opt.* **1996**, *35*, 2702–2710. [CrossRef]

42. Sun, B.; Roichman, Y.; Grier, D.G. Theory of holographic optical trapping. *Opt. Express* **2008**, *16*, 15765–15776. [CrossRef]

43. Waterman, P.C. Matrix formulation of electromagnetic scattering. *Proc. IEEE* **1965**, *53*, 805–812. [CrossRef]

44. Qi, X.; Nieminen, T.A.; Stilgoe, A.B.; Loke, V.L.Y.; Rubinsztein-Dunlop, H. Comparison of T-matrix calculation methods for scattering by cylinders in optical tweezers. *Opt. Lett* **2014**, *39*, 4827–4830. [CrossRef]

45. Nieminen, T.A.; Rubinsztein-Dunlop, H.; Heckenberg, N.R.; Bishop, A.I. Numerical modelling of optical trapping. *Comput. Phys. Commun.* **2001**, *142*, 468–471. [CrossRef]

46. Nieminen, T.A.; Rubinsztein-DUnlop, H.; Heckenberg, N.R. Calculation of the T-matrix: General considerations and application of the point-matching method. *J. Quant. Spectrosc. Radiat.* **2003**, *79*, 1019–1029. [CrossRef]

47. Loke, V.L.Y.; Nieminen, T.A.; Heckenberg, N.R.; Rubinsztein-Dunlop, H. T-matrix calculation via discrete dipole approximation, point matching and exploiting symmetry. *J. Quant. Spectrosc. Radiat.* **2009**, *110*, 1460–1471. [CrossRef]

48. Nieminen, T.A.; Loke, V.L.Y.; Stilgoe, A.B.; Heckenberg, N.R.; Rubinsztein-Dunlop, H. T-matrix method for modelling optical tweezers. *J. Mod. Opt.* **2011**, *58*, 528–544. [CrossRef]

49. Gouesbet, G. T-matrix formulation and generalized Lorenz-Mie theories in spherical coordinates. *Opt. Commun.* **2010**, *283*, 517–521. [CrossRef]

50. Nieminen, T.A.; du Preez-Wilkinson, N.; Stilgoe, A.B.; Loke, V.L.Y.; Bui, A.A.M.; Rubinsztein-Dunlop, H. Optical tweezers: Theory and modelling. *J. Quant. Spectrosc. Radiat. Transf.* **2014**, *146*, 59–80. [CrossRef]

51. Svoboda, K.; Schmidt, C.F.; Schnapp, B.J.; Block, S.M. Direct observation of kinesin stepping by optical trapping interferometry. *Nature* **1993**, *365*, 721–727. [CrossRef]

52. Ghislain, L.P.; Webb, W.W. Scanning-force microscope based on an optical trap. *Opt. Lett.* **1993**, *18*, 1678–1680. [CrossRef]

53. Bui, A.A.M.; Kashchuk, A.V.; Balanant, M.A.; Nieminen, T.A.; Rubinsztein-Dunlop, H.; Stilgoe, A.B. Calibration of force detection for arbitrarily shaped particles in optical tweezers. *Sci. Rep.* **2018**, *8*, 10798:1–10798:12. [CrossRef]

54. Smith, S.B.; Cui, Y.; Bustamante, C. Overstretching B-DNA: The Elastic Response of Individual Double-Stranded and Single-Stranded DNA Molecules. *Science* **1996**, *271*, 795–799.. [CrossRef]

55. Berg-Sørensen, K.; Flyvbjerg, H. Power spectrum analysis for optical tweezers. *Rev. Sci. Instrum.* **2004**, *75*, 594–612. [CrossRef]

56. Smith, S.B.; Cui, Y.; Bustamante, C. [7] Optical-trap force transducer that operates by direct measurement of light momentum. *Methods Enzymol.* **2003**, *361*, 134–162. [CrossRef]

57. Farr'e, A.; Montes-Usategui, M. A force detection technique for single-beam optical traps based on direct measurement of light momentum changes. *Opt. Express* **2010**, *18* 11955–11968. [CrossRef] [PubMed]

58. Ashkin, A.; Dziedzic, J.M.; Yamane, T. Optical trapping and manipulation of single cells using infrared laser beams. *Nature* **1987**, *330*, 769–771. [CrossRef] [PubMed]

59. Svoboda, K.; Block, S.M. Biological Applications of Optical Forces. *Annu. Rev. Biophys.* **1994**, *23*, 247–285. [CrossRef] [PubMed]

60. Mills, J.; Qie, L.; Dao, M.; Lim, C.; Suresh, S. Nonlinear elastic and viscoelastic deformation of the human red blood cell with optical tweezers. *Mech. Chem. Biosyst.* **2004**, *1*, 169–180. [CrossRef]

61. Capitanio, M.; Pavone, F.S. Interrogating Biology with Force: Single Molecule High-Resolution Measurements with Optical Tweezers. *Biophys. J.* **2013**, *105*, 1293–1303. [CrossRef]

62. Grigaravicius, P.; Greulich, K.O.; Monajembashi, S. Laser Microbeams and Optical Tweezers in Ageing Research. *ChemPhysChem* **2009**, *10*, 79–85. [CrossRef]

63. Suei, S.; Raudsepp, A.; Kent, L.M.; Keen, S.A.J.; Filichev, V.V.; Williams, M.A.K. DNA visualization in single molecule studies carried out with optical tweezers: Covalent versus non-covalent attachment of fluorophores. *Biochem. Biophys. Res. Commun.* **2015**, *466*, 226–231. [CrossRef]

64. Sischka, A.; Toensing, K.; Eckel, R.; Wilking, S.D.; Sewald, N.; Ros, R.; Anselmetto, D. Molecular mechanisms and kinetics between DNA and DNA binding ligands. *Biophys. J.* **2005**, *88*, 404–411. [CrossRef]

65. McCauley, M.J.; Wiliams, M.C. Mechanisms of DNA binding determined in optical tweezers experiments. *Biopolymers* **2007**, *85*, 154–168. [CrossRef]

66. Smith, S.B.; Finzi, L.; Bustamante, C. Direct mechanical measurements of the elasticity of single DNA molecules by using magnetic beads. *Science* **1992**, *258*, 1122–1126. [CrossRef]

67. Simmons, R.M.; Finer, J.T.; Chu, S.; Spudich, J.A. Quantitative measurements of force and displacement using an optical trap. *Biophys. J.* **1996**, *70*, 1813–1822. [CrossRef]

68. Lien, C.-H.; Wei, M.-T.; Tseng, T.-Y.; Lee, C.-D.; Wang, C.; Wang, T.-F.; Ou-Yang, H.D.; Chiou, A. Probing the dynamic differential stiffness of dsDNA interacting with RecA in the enthalpic regime. *Opt. Express* **2009**, *17*, 20376–20385. [CrossRef] [PubMed]

69. Bouchiat, C.; Wang, M.D.; Allemand, J.F.; Strick, T.; Block, S.M.; Croquette, V. Estimating the persistance length of a worm-like chain molecule from force-extension measurements. *Biophys. J.* **1999**, *76*, 409–413. [CrossRef]

70. Wang, M.D.; Yin, H.; Landick, R.; Gelles, J.; Block, S.M. Stretching DNA with optical tweezers. *Biophys. J.* **1997**, *72*, 1335–1346. [CrossRef]

71. Bustamante, C.; Marko, J.F.; Siggia, E.D.; Smith, S. Entropic elasticity of lambda-phage DNA. *Science* **1994**, *265*, 1599–1600. [CrossRef]

72. Marko, J.F.; Siggia, E.D. Stretching DNA. *Macromolecules* **1995**, *28*, 8759–8770. [CrossRef]

73. Shivashankar, G.V.; Feingold, M.; Krichevsky, O.; Libchaber, A. RecA polymerization on double-stranded DNA by using single-molecule manipulation: The role of ATP hydrolysis. *Proc. Natl. Acad. Sci. USA* **1999**, *96*, 7916–7921. [CrossRef]

74. van Mameren, J.; Gross, P.; Farge, G.; Hoojiman, P.; Modesti, M.; Falkenberg, M.; Wuite, G.J.L.; Peterman, E.J.G. Unraveling the structure of DNA during overstretching by using multicolor, single-molecule fluorescence imaging. *Proc. Natl. Acad. Sci. USA* **2009**, *106*, 18231–18236. [CrossRef]

75. Heller, I.; Hoekstra, T.P.; King, G.A.; Peterman, E.J.G.; Wuite, G.J.L. Optical tweezers analysis of DNA-protein complexes. *Chem. Rev.* **2014**, *114*, 3087–3119. [CrossRef]

76. Mohanty, S.K.; Rapp, A.; Monajembashi, S.; Gupta, P.K.; Greulich, K.O. Comet assay measurements of DNA damage in cells by laser microbeams and trapping beams with wavelengths spanning a ramge of 308 nm to 1064 nm. *Radiat. Res.* **2002**, *157*, 378–385.[0378:CAMODD]2.0.CO;2. [CrossRef]

77. Liang, H.; Vu, K.T.; Krishnan, P.; Trang, T.C.; Shin, D.; Kimel, S.; Berns, M.W. Wavelength dependence of cell cloning efficiency after optical trapping. *Biophys. J.* **1996**, *70*, 1529–1533. [CrossRef]

78. Neuman, K.C.; Chadd, E.H.; Liou, G.F.; Bergman, K.; Block, S.M. Characterization of photodamage to escherichia coli in optical traps. *Biophys. J.* **1999**, *77*, 2856–2863. [CrossRef]

79. Landry, M.P.; McCall, P.M.; Qi, Z.; Chemla, Y.R. Characterization of photoactivated singlet oxygen damage in single-molecule optical trap experiments. *Biophys. J.* **2009**, *97*, 2128–2136. [CrossRef] [PubMed]

80. Bl'azquez-Castro, A. Optical Tweezers: Phototoxity and Thermal Stress in Cells and Biomolecules. *Micromachines* **2019**, *10*, 507:1–507:42. [CrossRef]

81. Liu, Y.; Cheng, D.K.; Sonek, G.J.; Berns, M.W.; Chapman, C.F.; Tromberg, B.J. Evidence for localized cell heating induced by infrared optical tweezers. *Biophys. J.* **1995**, *68*, 2137–2144. [CrossRef]

82. Peterman, E.J.G.; Gittes, F.; Schmidt, C.F. Laser-induced heating in optical traps. *Biophys. J.* **2003**, *84*, 1308–1316. [CrossRef]

83. Nagar, H.; Admon, T.; Goldman, D.; Eyal, A.; Roichman, Y. Optical trapping below the diffraction limit with a tunable beam waist using super-oscillating beams. *Opt. Lett.* **2019**, *44*, 2430–2433. [CrossRef]

84. Jeffries, G.D.M.; Edgars, J.S.; Zhao, Y.; Shelby, J.P.; Fong, C.; Chiu, D.T. Using polarization-shaped optical vortex traps for single-cell nanosurgery. *Nano Lett.* **2007**, *7*, 415–420. [CrossRef]

85. Kairdolf, B.A.; Smith, A.M.; Stokes, T.H.; Wang, M.D.; Young, A.N.; Nie, S. Semiconductor quantum dots for bioimaging and biodiagnostic applications. *Annu. Rev. Anal. Chem.* **2013**, *6*, 143–162. [CrossRef]

86. Chowdhury, A.; Waghmare, D.; Dasgupta, R.; Majumder, S.K. Red blood cell membrane damage by light-induced thermal gradient under optical trap. *J. Biophotonics* **2018**, *11*, 201700222:1–201700222:10. [CrossRef]

87. König, K.; Tadir, Y.; Patrizio, P.; Berns, M.W.; Tromberg, B.J. Effects of ultraviolet exposure and near infrared laser tweezers on human spermatozoa. *Hum. Reprod.* **1996**, *11*, 2162–2164. [CrossRef] [PubMed]

88. Liu, Z.; Wang, L.; Zhang, Y.; Liu, C.; Wu, J.; Zhang, Y.; Yang, X.; Zhang, J.; Yang, J.; Yuam, L. Optical funnel for living cells trap. *Opt. Commun.* **2019**, *431*, 196–198. [CrossRef]

89. Shao, B.; Shi, L.Z.; Nascimento, J.M.; Botvinick, E.L.; Ozkan, M.; Berns, M.W.; Esener, S.C. High-throughput sorting and analysis of human sperm with a ring-shaped laser trap. *Biomed. Microdevices* **2007**, *9*, 361–369. [CrossRef] [PubMed]

90. Cizmar, T.; Siler, M.; Zemanek, P.; An optical nanotrap array movable over a milimetre range. *Appl. Phys.* **2006**, *84*, 197–203. [CrossRef]

91. Shao, B.; Esener, S.C.; Nascimento, J.M.; Berns, M.W.; Botvinivk, E.L.; Ozkan, M. Size tunable three-dimensional annular laser trap based on axicons. *Opt. Lett.* **2006**, *31*, 3375–3377. [CrossRef] [PubMed]

92. Manek, I.; Ovchinnikov, Y.B.; Grimm, R. Generation of a hollow laser beam for atom trapping using an axicon. *Opt. Commun.* **1998**, *147*, 67–70. [CrossRef]

93. Arlt, J.; Dholakia, K.; Soneson, J.; Wright, E.M. Optical dipole traps and atomic waveguides based on Bessel light beams. *Phys. Rev. A* **2001**, *63*, 063602:1–063602:8. [CrossRef]

94. Dao, M.; Lim, C.T.; Suresh, S. Mechanics of the human red blood cell deformed by optical tweezers. *J. Mech. Phys. Solids* **2003**, *51*, 2259–2280. [CrossRef]

95. Pollard, M.R.; Botchway, S.W.; Chichkov, B.; Freeman, E.; Halsall, R.N.J.; Jenkins, D.W.K.; Loader, I.; Ovsianikov, A.; Parker, A.W.; Stevens, R.; et al. Optically trapped probes with nanometer-scale tips for femto-Newton force measurement. *New J. Phys.* **2010**, *12*, 113056:1–113056:15. [CrossRef]

96. Phillips, D.B.; Simpson, S.H.; Grieve, J.A.; Bowman, R.; Gibson, G.M.; Padgett, M.J.; Rarity, J.G.; Hanna, S.; Miles, M.J.; Carberry, D.M. Force sensing with a shaped dielectric micro-tool. *EPL* **2012**, *99*, 58004:1–58004:6. [CrossRef]

97. Avci, E.; Ohara, K.; Nguyen, C.N.; Theeravithayangkura, C.; Kojima, M.; Tanikawa, T.; Mae, Y.; Arai, T. High-Speed Automated Manipulation of Microobjects Using a Two-Fingered Microhand. *IEEE Trans. Ind. Electron.* **2015**, *62*, 1070–1079. [CrossRef]

98. Mittas, A.; Dickey, F.M.; Holswade, S.C. Modeling an optical micromachine probe. In Proceedings of the Conference: Annual meeting of the Society of Photo-Optical Instrumentation Engineers, San Diego, CA, USA, 27 July–1 August 1997.

99. Rodrigo, P.J.; Gammelgaard, L.; Bøggild, P.; Perch-Nielsen, I.R.; Glückstad, J. Actuation of microfabricated tools using multiple GPC-based counterpropagating-beam traps. *Opt. Express* **2005**, *13*, 6899–6904. [CrossRef]

100. Probst, M.; Hürzeler, C.; Borer, R.; Nelson, B.J. A Microassembly System for the Flexible Assembly of Hybrid Robotic Mems Devices. *Int. J. Optomechatron.* **2009**, *3*, 69–90. [CrossRef]

101. Gerratt, A.P.; Penskiy, I.; Bergbreiter, S. Integrated silicon-PDMS process for microrobot mechanisms. In Proceedings of the 2010 IEEE International Conference on Robotics and Automation, Anchorage, AK, USA, 3–7 May 2010; pp. 3153–3158.

102. Hull, C.W. Apparatus for Production of Three-Dimensional Objects by Stereolithography. U.S. Patent US4575330A, 17 August 1986.

103. Galajda, P.; Ormos, P. Complex micromachines produced and driven by light. *Appl. Phys. Lett.* **2001**, *78*, 249–251. [CrossRef]

104. Göppert-Mayer, M. Über Elementarakte mit zwei Quantensprüngen. *Ann. Phys.* **1931**, *401*, 273–294. [CrossRef]

105. La Porta, A.; Offrein, B.J.; Soganci, I.M. Method of Manufacturing a Three Dimensional Photonic Device by Two Photon Absorption Polymerization. U.S. Pantent US13,898,869, 7 April 2015.

106. Hohmann, J.K.; Renner, M.; Waller, E.H.; von Freymann, G. Three-Dimensional μ-Printing: An Enabling Technology. *Adv. Opt. Mater.* **2015**, *3*, 1488–1507. [CrossRef]

107. Ma, J.; Cheng, W.; Zhang, S.; Feng, D.; Jia, T.; Sun, Z.; Qiu, J. Coherent quantum control of two-photon absorption and polymerization by shaped ultrashort laser pulses. *Laser Phys. Lett.* **2013**, *10*, 085304:1–085304:7. [CrossRef]

108. Kim, J.M.; Muramatsu, H. Two-Photon Photopolymerized Tips for Adhesion-Free Scanning-Probe Microscopy. *Nano Lett.* **2005**, *5*, 309–314. [CrossRef]

109. Kawata, S.; Sun, H.-B.; Tanaka, T.; Takada, K. Finer features for functional microdevices. *Nature* **2001**, *412*, 697–698. [CrossRef]

155. Constable, A.; Kim, J.; Mervis, J.; Zarinetchi, F.; Prentiss, M. Demonstration of a fiber-optical light-force trap. *Opt. Lett.* **1993**, *18*, 1867–1869. [CrossRef]

156. Zhao, X.; Zhao, N.; Shi, Y.; Xin, H.; Li, B. Optical Fiber Tweezers: A Versatile Tool for Optical Trapping and Manipulation. *Micromachines* **2020**, *11*, 114. [CrossRef]

157. Oh, M.-J.; Kuhr, F.; Byfield, F.; Levitan, I. Micropipette aspiration of substrate-attached cells to estimate cell stiffness. *J. Vis. Exp.* **2012**, *67*, e3886. [CrossRef] [PubMed]

158. Tanaka, Y. Double-arm optical tweezer system for precise and dexterous handling of micro-objects in 3D workspace. *Opt. Lasers Eng.* **2018**, *111*, 65–70. [CrossRef]

159. Bianchi, S.; Di Leonardo, R. Real-time optical micro-manipulation using optimized holograms generated on the GPU. *Comput. Phys. Commun.* **2010**, *181*, 1444–1448. [CrossRef]

Review

Optical Fiber Tweezers: A Versatile Tool for Optical Trapping and Manipulation

Xiaoting Zhao [†], Nan Zhao [†], Yang Shi, Hongbao Xin * and Baojun Li

Institute of Nanophotonics, Jinan University, Guangzhou 511443, China; xiaotingzhao@stu2019.jnu.edu.cn (X.Z.); zhaonan@stu2018.jnu.edu.cn (N.Z.); shiyang@jnu.edu.cn (Y.S.); baojunli@jnu.edu.cn (B.L.)
* Correspondence: hongbaoxin@jnu.edu.cn; Tel.: +86-20-3733-6704
† These authors contributed equally to this work.

Received: 7 December 2019; Accepted: 16 January 2020; Published: 21 January 2020

Abstract: Optical trapping is widely used in different areas, ranging from biomedical applications, to physics and material sciences. In recent years, optical fiber tweezers have attracted significant attention in the field of optical trapping due to their flexible manipulation, compact structure, and easy fabrication. As a versatile tool for optical trapping and manipulation, optical fiber tweezers can be used to trap, manipulate, arrange, and assemble tiny objects. Here, we review the optical fiber tweezers-based trapping and manipulation, including dual fiber tweezers for trapping and manipulation, single fiber tweezers for trapping and single cell analysis, optical fiber tweezers for cell assembly, structured optical fiber for enhanced trapping and manipulation, subwavelength optical fiber wire for evanescent fields-based trapping and delivery, and photothermal trapping, assembly, and manipulation.

Keywords: optical fiber tweezers; optical trapping and manipulation; optical force; photothermal effect; evanescent fields; cell trapping and assembly

---

1. Introduction

Optical forces have been widely used for optical trapping and manipulation of particles while using laser beams since 1970, when Arthur Ashkin used two counter-propagating and continuous wave focused beams to trap particle [1]. In 1986, Ashkin and co-workers used a single tightly focused laser beam, which realized the stable trapping of particles. They then named the optical-trapping technique as optical tweezers [2], which we refer to conventional optical tweezers (COTs). Over the years that followed, Ashkin and co-workers carried out a series of studies, which not only realized the capture of particles ranging from tens of nanometers to tens of microns with a single focused beam, but also realized the capture of viruses and bacteria [2,3].

After nearly 50 years of development, optical trapping and manipulation via COTs have made great progresses in both methods and applications, with manipulated samples ranging from different dielectric particles to biological cells and biomolecules [4,5]. However, a high numerical aperture (NA) objective is necessary for the light focusing via COTs. In addition, different optical components for beam expanding and steering are also essential. This bulky structure makes it lack manipulation and control flexibility. Alternatively, in 1998, holographic optical tweezers (HOTs) that can realize multiple traps with complex structured light fields have been developed [6]. These multiple traps can be achieved by means of computer generated holograms via the modulation of spatial light modulators. Such HOTs greatly increase the manipulation and control degrees, and they are also widely used for multiple particle trapping and manipulation [7–10]. However, it is difficult for the stable trapping of particles in the nanometer scale due to the diffraction limit. In the late 2000s, surface plasmon-based optical tweezers (SPOTs) [11,12] have been created, which can realize the stable

---

trapping of nanoparticles, even single molecules with a few nanometer scale, to realize the nanoscale trapping and manipulation [13,14]. Though with the ability for the trapping of nanoscale particles, the SPOTs necessitate carefully designed and elaborate nanostructures. These nanostructures also limit the manipulation flexibility. Techniques, such as COTs, HOTs, and SPOT, involve complex devices and components with inflexible control. Therefore, it is particularly important to seek a simple and flexible control and manipulation tool. The development of optical fiber tweezers (OFTs) [15,16] makes it a versatile candidate for optical trapping and the manipulation of different samples. OFTs possess exceptional advantages in the manipulation flexibility due to the simple structure with only optical fibers. The fiber can be inserted into thick samples and turbid media, which greatly increases the sample applicability. In addition, OFTs make optical manipulation a low cost technique due to the easy fabrication procedures. OFTs can also be integrated into small devices, such as optofluidic channels and chips [17]. OFTs were first presented in 1993, where two aligned single-mode optical fibers were used for optical trapping [18]. While tiny particles and cells can be directly captured and manipulated using the optical scattering force generated by two fibers [19], the manipulation flexibility is limited by the two fibers. Indeed, a single optical fiber can also be used for particle trapping and manipulation. In 1997, the first optical manipulation using a single tapered optical fiber was reported [20]. These single fiber-based OFTs greatly increase the manipulation flexibility. The end of a single fiber for light focusing is similar to that in COTs after being drawn into a lenticular shape, which creates a stronger gradient force on the particle and makes it easier for optical trapping [21–23]. Both dual and single fibers can be used for stable trapping, as shown in Figure 1.

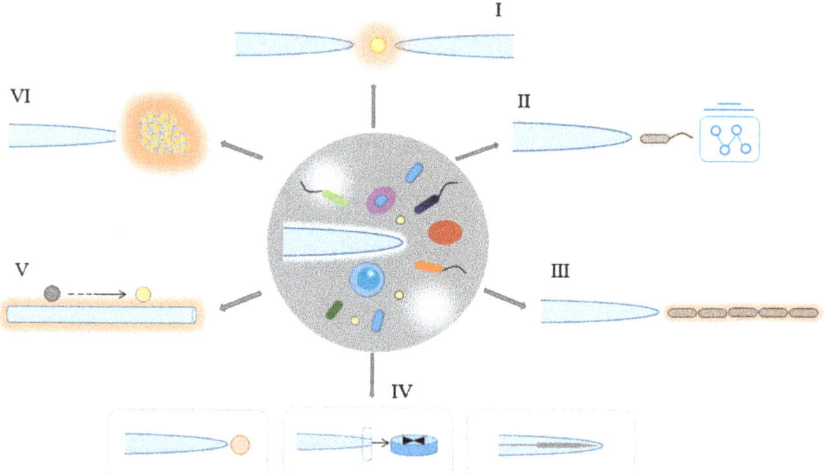

Figure 1. The overview figure of optical fiber tweezers (OFTs) for particle and cell trapping and manipulation. (**I**) Dual fiber tweezers for trapping and manipulation; single optical fiber tweezers for (**II**) cell trapping and single cell analysis, and (**III**) cell assembly; (**IV**) structured optical fiber tweezers for trapping and manipulation; (**V**) subwavelength optical fiber wire for evanescent fields-based trapping and delivery; and, (**VI**) optical fiber-based massive photothermal assembly and manipulation.

OFTs can be divided into two categories, according to their working principles: photothermal effect and optical force. The former generates trapping force by laser-induced thermal and acoustic gradients in the surrounding medium of the particles [24], while the latter has been traditionally decomposed into two components: the scattering force in the direction of light propagation and the gradient force in the direction of the optical intensity gradient [25].

For the purpose of this review, we will focus on the OFTs and summarize their recent advances in the trapping and manipulating of tiny objects, especially for cells.

2. Theoretical Analysis of Optical Forces Exerted on Particles

Optical force will be exerted on the particles near the focal point by the momentum transfer from the scattering of incident photons when a high intensity laser beam is irradiated on dielectric particles. The resulting optical force traditionally consists of two components: a scattering force and a gradient force. The Rayleigh scattering condition will be satisfied when the wavelength of laser beam is much longer than the size of trapped particle [25]. With the above condition, the optical force can be obtained by treating particles as point dipoles. For a particle of radius a, this force is given by

$$F_{\text{scatt}} = \frac{I_0 \sigma n_{\text{m}}}{c} \tag{1}$$

$$\text{With } \sigma = \frac{128\pi^5 a^6}{3\lambda^4}\left(\frac{m^2-1}{m^2+2}\right)^2 \tag{2}$$

where I_0 is the intensity of the incident light, σ is the scattering cross section of the particle, n_{m} is the refractive index of the surrounding medium, c is the speed of light in vacuum, m is the ratio of the refractive index of the particle to that of the medium (n_p/n_{m}), and λ is the wavelength of the trapping laser. The scattering force is in the same direction as the incident light and it is proportional to the intensity of the light. The optical gradient force is generated by the interaction between the induced dipole and the non-uniform field, as given by

$$F_{\text{grad}} = \frac{2\pi\alpha}{cn_{\text{m}}^2}\nabla I_0 \tag{3}$$

$$\text{with } \alpha = n_{\text{m}}^2 a^3\left(\frac{m^2-1}{m^2+2}\right) \tag{4}$$

However, for many cases, the size of the trapped particles is comparable to the wavelength of the trapping laser beam. In this situation, the point-dipole approach is invalid. More complete electromagnetic theories are necessary for the force calculation. Fortunately, numerical simulation and calculation methods that are based on electromagnetic theories, such as the finite element method and finite difference time domain method, can be used for the calculation of optical forces. The total optical force (F_O) exerted on the particle can be calculated by calculating the integral of the time-independent Maxwell stress tensor ($<\mathbf{T_M}>$) along the total external surface of the particle. $<\mathbf{T_M}>$ can be expressed as

$$\langle \mathbf{T_M} \rangle = \mathbf{DE^*} + \mathbf{HB^*} - 1/2(\mathbf{D \cdot E^*} + \mathbf{HB^*})\mathbf{I} \tag{5}$$

where $\mathbf{D}$ and $\mathbf{H}$ are the electric displacement and magnetic field, respectively; $\mathbf{E^*}$ and $\mathbf{B^*}$ are the complex conjugates of the electric field $\mathbf{E}$ and magnetic flux field $\mathbf{B}$, respectively; and, $\mathbf{I}$ is the isotropic tensor. F_O can be expressed as

$$\mathbf{F_O} = \oint_s (\langle \mathbf{T_M} \rangle \cdot \mathbf{n})d\mathbf{S} \tag{6}$$

where $\mathbf{n}$ is the surface normal vector.

3. Dual Fiber Tweezers for Trapping and Manipulation

The dual fiber tweezers mainly use the scattering force to trap particles cooperatively by two optical fibers. It can form a large enveloping area and easily accommodate large objects due to its beam divergence. In addition, the dual fiber tweezers do not use focused light, resulting in minimal radiation damage to living cells [26]. Jess et al. used a fiber arm with an output power of 800 mW to trap a 100 μm polymer sphere [27] (Figure 2a). Decombe et al. achieved the capture of a 1 μm polystyrene sphere at an optical power of 2 mW by using optical tweezers with the tips of two chemically corroded fibers [28].

The rotation of the trapped particles can also be achieved by the dual fiber tweezers. For example, a fiber spanner tool [29], which was formed by two transversely offset fibers with counterpropagating laser beams that can rotate particles, like a spanner tool, was used to introduce the lateral offset between the two back-propagating divergent beams from the single-mode fiber. The center of the fiber is laterally biased and the optical fiber is used to trap and rotate the human smooth muscle cell (hSMC) (Figure 2b). If there is a zero offset, the object can remain in a static state. The object begins to rotate when two fibers have a relative displacement [29]. The rotation of living cells can be performed in a dual-beam optical fiber trap integrated into a modular laboratory chip system [30].

Alternatively, the cells can be stretched after being captured with a double beam. Guck et al. constructed a device and named it an optical stretcher [31]. The principle of the optical stretcher is as follows: when placing the dielectric object in the middle of two opposed, non-focused laser beams, the total force acting on the object is zero, but the surface forces are additive, which causes the object to be stretched along the beam axis. As shown in Figure 2c, Müller cell can be trapped, aligned, and stretched out by two counter-propagating near-infrared laser beams diverging from the optical fibers, where the incident laser wavelength is 1064 nm [32]. The cell stretching results from the additive surface force on the cell membrane due to momentum changes of photons [31]. The red blood cells can also be stretched using dual-trap optical tweezers after placing them in the suspension and expanding into a sphere [33].

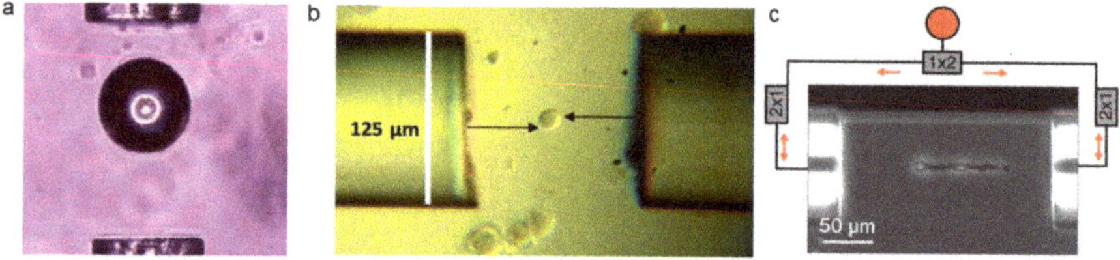

Figure 2. Dual fiber tweezers for cell trapping and manipulation. (**a**) A 100 µm polymer is trapped by the dual fibers. Adapted with permission from Jess et al. [27]. (**b**) The human smooth muscle cell is captured and rotated by optical fiber at the center of two horizontally biased fibers (20 mW in each arm). Adapted with permission from Black et al. [29]. (**c**) The Müller cell is captured, aligned, and stretched by two near-infrared laser beams that propagate backwards from the fiber. Adapted with permission from Franze et al. [32].

The use of dual fiber tweezers also allows for more precise control of the nanoscale particles in addition to the manipulation of particles and cells with a relatively large size (micrometer scale). Figure 3a shows the schematic depiction of a dual-fiber-nanotip method for the manipulation of multiwalled carbon nanotubes (MWCNTs) with an outer diameter of only 50 nm and a length of 0.9 µm. The position and orientation of the MWCNT can be adjusted by changing the distance between the two nanotips and the input power of each nanotip. When P_1 (optical power from the left fiber) remains unchanged and P_2 (optical power from the right fiber) power increases, the orientation and position of the MWCNT change accordingly [34]. Xu et al. realized the three-dimensional (3D) optical trapping of silver nanoparticles and nanowires while using dual focused coherent beams [35] (Figure 3c). As the distance d (the gap between two fiber probes, FPs) increases, the initial stable trap state of the silver nanowires (diameter: 330 nm, length: 2.1 µm) is destroyed and the transverse Poynting vector points outward, which causes the particle to escape (Figure 3d-I,II). The nanowires are rotated clockwise by increasing the transverse distance between FP1 and FP2 to form an asymmetric field distribution (Figure 3e). Additionally, Liu et al. proved that the inclined dual-fiber optical tweezers have the ability of 3D trapping, which is related to the inclination angle of the fiber [36]. The combination of 3D printed Fresnel lenses on dual fiber surface was reported to be an excellent method to increase the trapping efficiency and stability [37]. In this scenario, the trap stiffness is increased by a substantial factor of 35–50.

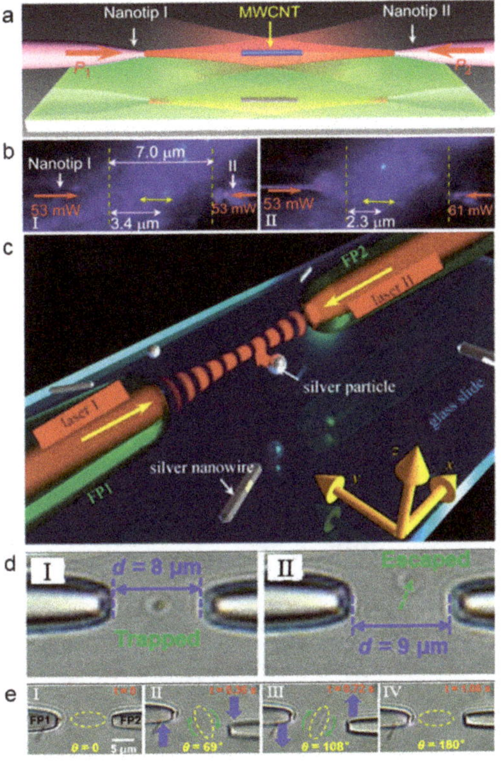

Figure 3. (**a**) Schematic depiction of the orientation and displacement of a single multiwalled carbon nanotubes (MWCNT) controlled using a dual-fiber-nanotip method. (**b**) Dark-field optical microscope images of MWCNT at different output power. Adapted with permission from Xin et al. [34]. (**c**) Schematic depiction of three-dimensional (3D) optical trapping of silver nanostructures. (**d**) Optical traps images at different distances. (**e**) Continuous microscopic images of silver nanowires trapped by rotation. Adapted with permission from Xu et al. [35].

4. Single Optical Fiber Tweezers for Trapping and Single Cell Analysis

4.1. Single Optical Fiber Tweezers for Particle/Cell Trapping

Dual fiber tweezers still lack simplicity, so it is very important to realize the optical trapping and manipulation based on a single fiber, i.e., single optical fiber tweezers. In general, single optical fiber tweezers are tweezers that resulted from a single optical fiber. Figure 4a shows the working principle of particle trapping while using a single tapered fibre probe (TFP) [38], where D_A indicates the axial distance (x direction) of the particles to the tip of the TFP and D_T indicates the transverse distance (in y direction) of a particle to the axis. Once a laser beam is launched into the TFP, the light emerging from the TFP after focusing will exert an optical force on the particles. Optical force consists of two parts—the gradient force F_g and the scattering force F_s. F_g tends to attract particles, while F_s pushes away, and the resultant force of them together dominates the motion of the particles. The optical force on the particle can be calculated by integrating the Maxwell stress tensor around the particle. Figure 4b-I shows the calculated force F_x exerted on the particle along the TFP axis as a function of D_A (i.e., $D_T = 0$). When the particle is close to the TFP tip ($D_A < 13$ μm), F_x is negative, which means that there is a trapping force towards the tip. In this region, the gradient force F_g dominates the motion of particles and the particle can be trapped. When the particle is away from the TFP tip, F_x is positive, which means that there is a driving force in the direction of light propagation. In this region, the scattering force F_s dominates the motion of particles and the particle can be pushed away. This different optical force-based manipulation mechanism results from the light distribution near the fiber tip. For a clearer description, Figure 4c shows the light distribution near the fiber tip [39]. It can be seen that the laser beam is focused at the fiber tip. At the tip of the fiber, the optical intensity reaches the highest and, therefore, there is a strong optical gradient force near the tip, which can be used for trapping. The gradient of intensity in the converging beam pulls small objects towards the focal point, while the radiation pressure tends to push them away along the optical axis. With gradient force being dominated, a particle can be trapped in three dimensions near the focal point by the optical gradient force [40].

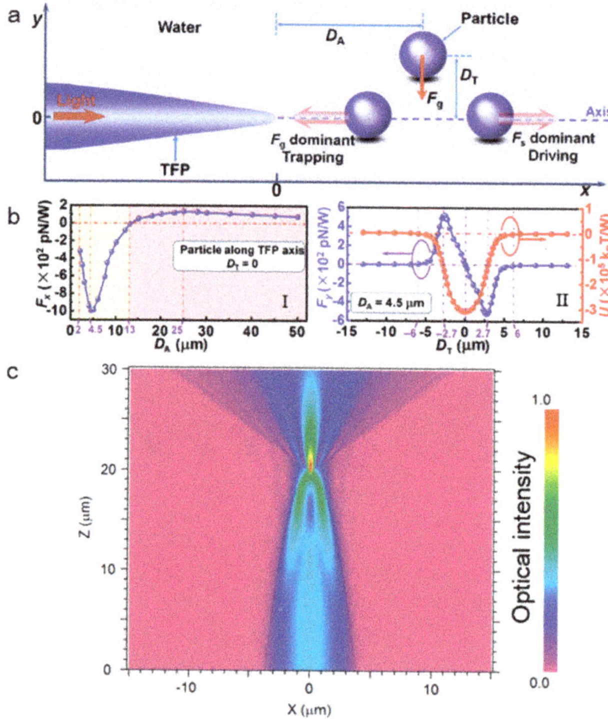

Figure 4. The principle for the single optical fiber tweezers trapping of particles. (**a**) Schematic illustration for particle trapping and manipulation with light launched into the tapered fibre probe (TFP). (**b**) Calculation result of the optical force: (I) axial force exerted on particles along the TFP axis as a function of D_A, (II) transverse force (F_y), and trapping potential (U) received by the particles at $D_A = 4.5$ μm. Adapted with permission from Xin et al. [38]. (**c**) Optical intensity distribution near the fiber tip. Adapted with permission from Liu et al. [39].

The requirement to trap a particle is the light focusing at the fiber end, so that an optical gradient force can be exerted on the particles near the fiber end. Additionally, the main effects are the optical forces. The trapping mechanism for optical fiber tweezers is the same as that in COTs, both are resulted from the optical force (optical gradient force for trapping). For the COTs, light is focused by a high NA objective and the particle is trapped near the focus by optical gradient force. In single OFTs, light is focused by the end surface of the fiber and the optical gradient force also traps the particle. The ability of OFTs to trap particles highly depends on the shape of the end of the fiber, which directly affects the light focusing at the fiber end. For an optical fiber with a flat end surface, the light is difficult to focus and, thus, the particle is difficult to be trapped. Instead, it will be pushed away by the optical scattering force. For a tapered optical fiber with a parabolic or convex surface, light will be focused at the fiber end, and particles can be trapped by the generated optical gradient force. Light is highly focused near the fiber tip, as shown in Figure 4c. This highly focused light will generate a strong optical gradient force for particle trapping. Actually, Xin et al. calculated the trapping efficiency of a 10-μm polystyrene particle in both OFTs and COTs [38]. The calculated efficiency is 0.227 and 0.12 for OFTs and COTs, respectively. That means that the trapping efficiency for OFTs is not weaker than COTs. The main difference between the FOTs and the COTs is that the COTs realize a truly non-contact trapping, while particle might be contacted with the fiber surface after trapping in OFTs. By changing the shape of tapered fiber end so that the focal point is a few microns away from the tip of the optical fiber, we can realize a non-contact optical trapping [41]. The fundamental limitation for the trapping is the diffraction limit. For particles within the diffraction limit, it is difficult to trap.

Optical fiber tweezers can be used to trap particles and cells of different sizes, which range from few hundreds of nanometers to tens of micrometers. Figure 5 shows the optical trapping of different particles/cells from small to large. The main influence factor is Brownian motion and diffraction limit for the trapping of small particles. As shown in Figure 5a, the SiO_2 sphere was 1.1 μm away from the optical fiber tip at $t_1 = 0\,s$, and it was trapped at the tip at $t_1 = 1\,s$. Likewise, *Escherichia coli* (*E. Coli*) was 1.3 μm away from the optical fiber tip at $t_1 = 0\,s$, and trapped at the tip at $t_1 = 1\,s$ [38]. Figure 5c shows a tapered optical fiber with abruptly tapered twin-core, which is fabricated by fusing and drawing the twin-core fiber [42]. The twin-core fiber extends the functionality of the fiber optical tweezers, and allows for the particle trapping with orientation. The single optical fiber tweezers can even be used for the trapping of cells larger than 10 μm. Figure 5d shows the trapping of a mammalian cell with a size of about 15 μm [43].

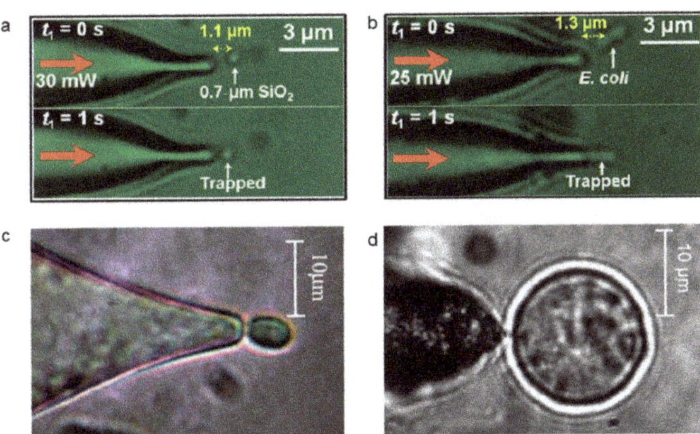

Figure 5. Single optical fiber tweezers are able to trap particles and cells with different sizes. Optical trapping of (**a**) a 0.7-μm-diameter SiO_2 particles. Adapted with permission from Xin et al. [38], (**b**) *E. coli*. Adapted with permission from Xin et al. [38], (**c**) yeast cell. Adapted with permission from Grier et al. [42], (**d**) Chinese hamster ovary cell. Adapted with permission from Mohanty et al. [43].

Single OFTs are widely used for optical trapping and manipulation due to the easy fabrication and simple structure. Without the need of high-NA objective for light focusing and the optical elements for laser beam expanding and light steering that are necessary in COTs, the OFTs have the advantages of compactness and high manipulation flexibility. By simply moving the fiber, the trapped particles can be flexibly manipulated accordingly. Figure 6 shows some examples for the flexible manipulation and moving of trapped particles and bacteria [38,44]. By simply moving the fiber, the trapped particles can be manipulated in three dimensions flexibly. Such manipulation enables the pick-up action for further precise arrangement of particles into designed patterns (Figure 6a). The manipulation can also be applied in flowing environment (Figure 6b,c) [44]. In addition, a single optical fiber can be inserted into the sample with any angles and any depth, which greatly increases the manipulation flexibility. Therefore, the single optical fiber tweezers can be used in many different environments, where COTs cannot be achieved. By cooperating with aspheric lens, optical trapping, and cooling of particles in vacuum has also been demonstrated [45]. For a direct comparison, Table 1 shows some features of single OFTs when compared with COTs.

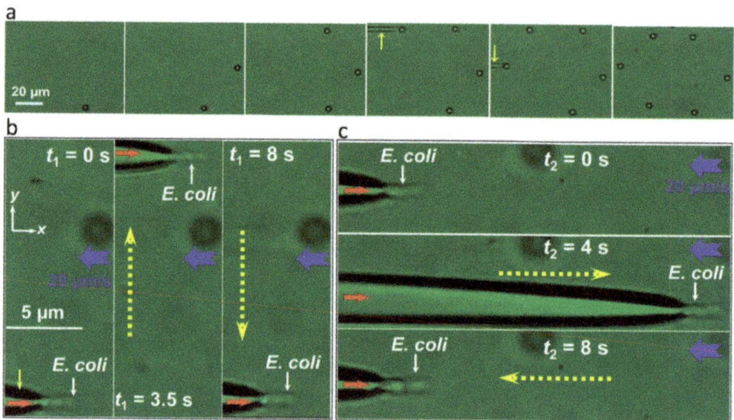

Figure 6. Images of using OFTs for flexible manipulation of trapped particles and bacteria. (**a**) Arrangement of six particles into a hexagon. The particles are picked up and placed into the designated postion after trapping by flexible moving the fiber (yellow arrow indicated). Adapted with permission from Xin et al. [38]. (**b,c**) Flexible manipulation of a trapped *E. coli* bacterium in flowing environment. The yellow solid arrow indicates the fiber, the blue arrows indicate the flowing direction. Adapted with permission from Xin et al. [44].

Table 1. Comparison of single optical fiber tweezers (OFTs) with the conventional optical tweezers (COTs).

Items	Single OFTs	COTs
Key components	Laser source, tapered optical fiber	Laser sourcehigh NA objectivea number of optical components for beam expanding and steering
Fabrication and construction	Easy, simply fabricate a tapered optical fiber with different methods	Carefully design the beam path via the adjusting of beam expanding and steering components are necessary
Integration capability	Highly compact, can be integrated into microfluidic platform	Not compact
Manipulation flexibility	Highly flexible, trapped particles can be delivered to any designated positions by simply moving the fiber	Less flexible, trapped particles can only be manipulated by controlling the focus through beam steering and modulation elements incorporated with the high-NA objective
Suspension applicability	Wide, the fiber can be inserted into suspensions with any different directions and depths for trapping and manipulation	Suspension depth and direction is limited due to the focus generated by the high-NA objective.

4.2. Single Optical Fiber Tweezers for Cell Analysis

In addition to the trapping of particles and cells, the single optical fiber tweezers are also widely used for single cell labelling and analysis after the cell is trapped. For example, a single bacterium can be labelled via the cotrapping of a single bacterium and upconversion nanoparticles (UCNPs) [46]. Figure 7a,b provide a clear description of how a tapered fiber probe (TF) traps a single UCNP and briefly labels the bacterium. A single UCNP ($\approx$120 nm) undergoing Brownian motion is trapped at the tip of the TF, with a 980 nm wavelength laser beam. It is difficult to observe uncaptured UCNPs because of the small size of UCNPs. When are UCNPs is trapped to the tip, it emits green light excited by the 980 nm laser and could, therefore, be directly observed. After the trapping of the UCNP, a bacterium can be further trapped, and another UCNP can be cotrapped at the end face of the bacterium. The emitting green light provides a direct way for the labeling and observation of a single bacterium that is otherwise impossible in dark environments. Figure 7b shows the images of labelled individual bacteria with different lengths. The labelling also allows for the single bacterium analysis via optical signals. Figure 7c shows the reflection signal of labeled bacteria with lengths of 1.8, 2.4, and 3.2 μm, respectively.

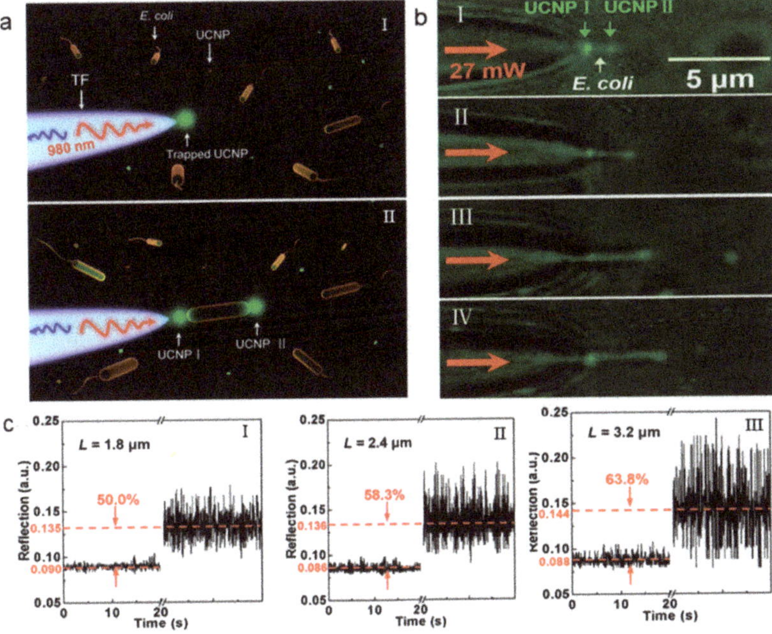

Figure 7. (a) Schematic and (b) experimental images of bacterial cotrapping and labelling. (c) Real-time reflected optical signal from the trapping and labeling of single bacteria with different sizes. Adapted with permission from Xin et al. [46].

In addition to the analysis of single bacteria from the optical signals, optical trapping via single OFTs also allows for the energy analysis of motile bacteria. Xin et al. used a modified tapered fiber to trap *E. coli* in solution and studied the kinematics of single bacteria [41]. Figure 8a shows the dynamic of a motile bacterium in the trapping potential. At t_1 = 1.7 s, the motile bacterium was stably trapped at a distance of 1.9 μm away from the fiber tip and kept halted for about 0.2 s (Figure 8a-I). At t_1 = 1.9 s, the bacteria began to struggle in the trapping potential after the stored energy release (Figure 8a-II). Later, it was trapped back toward the fiber (Figure 8a-II). At t_1 = 2.3 s, the bacterium was trapped back with a gap of 1.5 μm to the fiber tip. Figure 8b can analyze and explain the phenomenon of bacteria back-and-forth. In solution, the motile bacterium is swimming. When near the fiber tip, it was first trapped by the OFTs, and then halted for about 0.1–1 s. After that, the energy that was stored in the bacterium was released and converted into kinetic energy. If the trapping potential is smaller than bacterium energy, the bacterium can be released. Otherwise, the bacterium can be struggling in the trapping potential. Further, the bacterium can further escape from the trap if the stored energy is further released, and is larger than the trapping potential. Otherwise, the bacterium can be constantly trapped. This method provides a simple method for the analysis of bacterium dynamics.

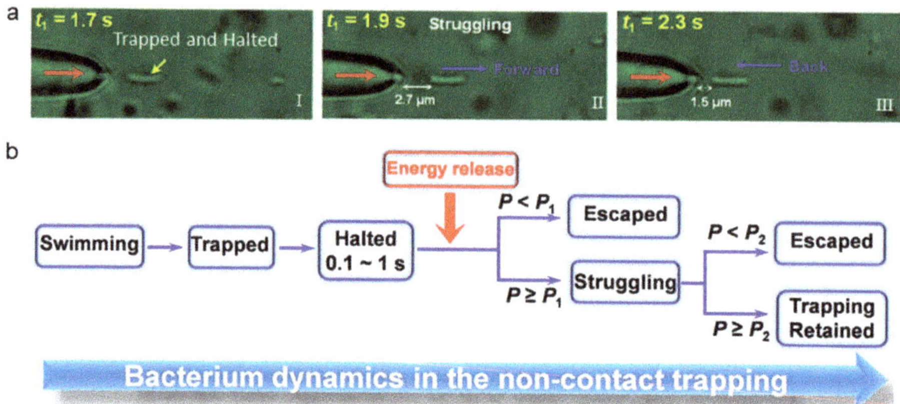

Figure 8. Dynamic analysis of motile bacteria via non-contact optical trapping of single bacteria. (**a**) Optical microscope images capture and struggle with the process of *E. coli* bacterium. (**b**) Schematic of the bacterium dynamics in the non-contact trapping. Adapted with permission from Xin et al. [41].

5. Optical Fiber Tweezers for Cell Assembly

In addition to the trapping and manipulation of a single particle, OFTs can also be used for the cell assembly, which is very important for the study of cell-cell interaction and communication. In addition, the assembly of randomly distributed cells into regular shaped structures and arrays also plays an important role in many other biomedical and bio-optical fields, such as tissue engineering [47], drug delivery, and targeted therapy [48,49].

5.1. Cell Assembly by the Optical Binding

Tam et al. used multiple optical fibers to form dense optical wells to capture and arrange parallel microspheres [50]. However, this approach is too cumbersome and it would be nice to capture and arrange particles with a single fiber. Figure 9a schematically shows the mechanism of particle chain formation while using a single optical fiber probe (TF) [51]. When a 980 nm laser beam is launched into the TF, the resulting optical force will act on the particles. Firstly, the scattering force will drive the first particle away along the TF axis. Subsequently, the particles beside the TF axis and near the first particle can be trapped to the TF axis resulting from transverse gradient force. Under the action of axial gradient force of the tail particle, the particle is closely bound to the former and, therefore, multiple particles and cells are assembled together by the cooperation of optical scattering force and gradient force. By using the above principle, the particles can be arranged into a one-dimensional particle chain (Figure 9b) or even a two-dimensional graphic particle array (Figure 9c). The binding ability also works on cells in the same manner. Figure 9d shows the yeast cell chain formed by optical binding. This method can even be used for the assembly of organelles inside a living cell. For example, Li et al. investigated the noncontact intracellular binding and controlled manipulation of chloroplasts in vivo while using optical fiber probes [52]. Figure 9e shows the optical binding of chloroplasts in plant cells using optical fiber probe (OFP), which is above a plant leaf with an approximately 3-μm gap between the fiber tip and the leaf. Figure 9f shows the arranged one-dimensional (1D) and two-dimensional (2D) chloroplast structures.

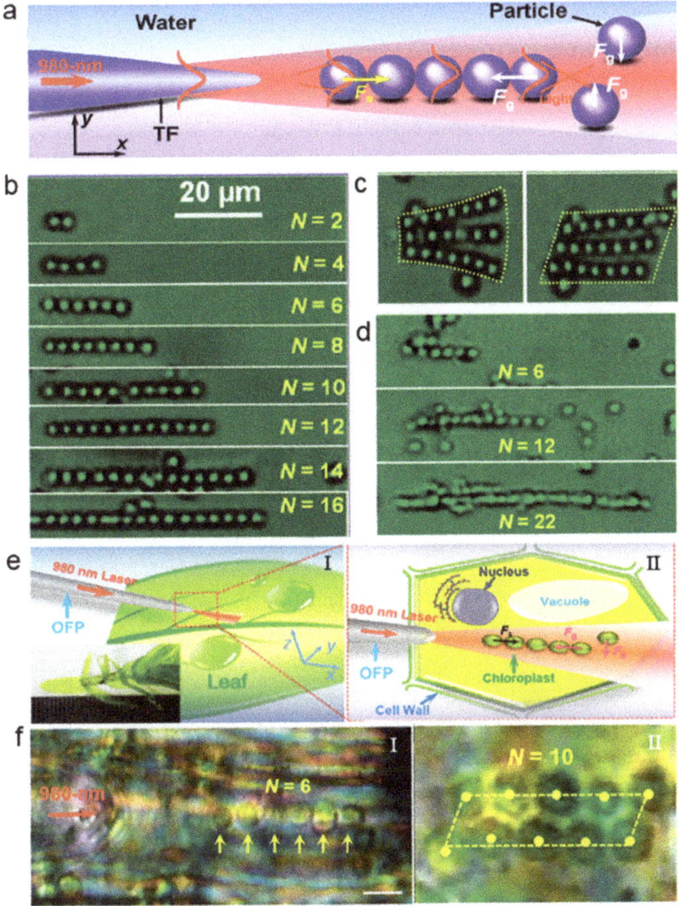

Figure 9. (**a**) Schematic of optical assembly of particles and cells via optical binding. Microscopic images of (**b**) one-dimensional (1D) patterned particle chains and (**c**) two-dimensional (2D) patterning of particle arrays. (**d**) Microscopic images of yeast cell chains. Adapted with permission from Xin et al. [51]. (**e-I**) Schematic of an optical fiber probe (OFP) which is aslant placed above the plant leaf with a laser at 980 nm launched. The inset shows a living plant (*Hydrilla verticillata*) on a glass slide. (**e-II**) Schematic of optical binding of chloroplasts inside a plant cell, which shows a row of chloroplasts confined in the optical axis and bound to each other, resulting from the cooperation of F_g and F_s. (**f**) Microscopic images of the arranged chloroplast chain and rectangle. Adapted with permission from Li et al. [52].

5.2. Cell Assembly by Extended Optical Gradient Force

Multiple cells can also be assembled solely by optical gradient force in addition to the cooperation between optical scattering force and gradient force for cell assembly. Xin et al. reported an extended optical gradient force-based method for cell assembly [53]. In general, a cell is first trapped at the tip of an ATF (abrupt tapered optical fiber) by optical gradient force, light can propagate along the cell, and refocused at the cell end surface, as shown in Figure 10a. This refocused light can generate an optical gradient force on other cells. In this regard, the optical gradient force is extended. Multiple cells are then assembled via the extended optical gradient force, and light can propagate along the multiple cells, Figure 10b shows the energy density distribution of the captured *E. coli* of (I) $N = 0$, (II) $N = 3$, (III) $N = 10$. It is can be seen that light is highly concentrated at the tip of the fiber, which provides a higher trapping efficiency to *E. coli*. Using this phenomenon, Xin et al. constructed biological optical waveguides (bio-WGs) based on *E. coli* bacteria, as shown in Figure 10c. This all cell-based bio-WGs opens the door for the fabrication of biological waveguides using cells. Because all of the materials of the bio-WGs are biological cells, these waveguides are highly biocompatible, when compared with conventional optical waveguides that are based on silica and other organic/inorganic materials. Based on this idea, Li et al. made a bio-nanospear in which the "handle" was made of a tapered fiber and the "head" was assembled with a yeast cell and a chain of nanosized *Lactobacillus acidophilus* (*L. acidophilus*) cells [54], as schematically shown in Figure 10d. At first, an 808 nm laser beam was launched into the tapered fiber to trap a yeast cell at the tip and then to trap a *L. acidophilus* behind the yeast. The spherical yeast focused the laser beam and exerted a strong optical force on the *L. acidophilus* cell that was trapped behind the yeast. Figure 10e is the image of the biological spear. With precise manipulation, the formed bio-nanospear can guide the input light to a specific location and detect optical signals from biological cells, such as individual leukemia cells in human blood. In addition, the group stably trapped a bio-microlens (yeast or a human cell) on a compact fiber probe by optical force, the excitation light was, therefore, confined in a subwavelength region, which resulted in the enhanced up-conversion fluorescence [55].

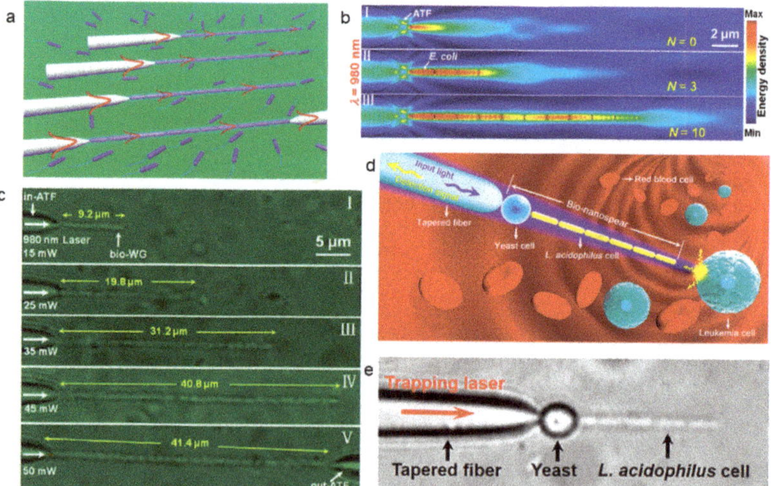

Figure 10. Optical assembly via extended optical gradient force. (**a**) Schematic of optical assembly and biological optical waveguides (bio-WGs) formation. (**b**) The distribution of energy density in ATF when different numbers of *E. coli* were captured (**c**) Images of formed bio-WGs with different lengths. Adapted with permission from Xin et al. [53]. (**d**) Schematic illustration of the assembled bio-nanospear. (**e**) Optical image of the bio-nanospear assembled from a yeast and *L. acidophilus* cells. Adapted with permission from Li et al. [54].

5.3. Cell Separation after Assembly

The separation and screening of particles can also be realized by OFTs after cell assembly. Liu et al. proposed a compact, miniaturized optofluidic chip integrated with OFT in a T-type microfluidic channel [56] (Figure 11a) for the selective capture of cells and bacteria. It was verified that the OFTs can capture and assemble *E. coli* cells, and it can drive red blood cells (RBCs), to experimentally verify the OFTs' selective capture capability. It is reported that the focusing characteristics of the fiber tip will change with its shape, which results in different optical forces. Taking advantage of this feature, a special fiber tip was used to provide push force to the red blood cells and trapping force to the *E. coli* cells, thus achieving cell sorting. Figure 11b experimentally verified the OFTs' selective trapping, which realized the trapping and delivery of *E. coli* and the push of RBCs. When the laser was turned off, the *E. coli* and RBCs were mixed together in the channel 1. Once laser was on, the *E. coli* were trapped and assembled, while the RBCs were immediately pushed away (Figure 11b-I). Finally, the trapped and assembled *E. coli* were sent to channel 2, while the RBCs were left in channel 1, achieving the 100% pure separation (Figure 11b-II–IV).

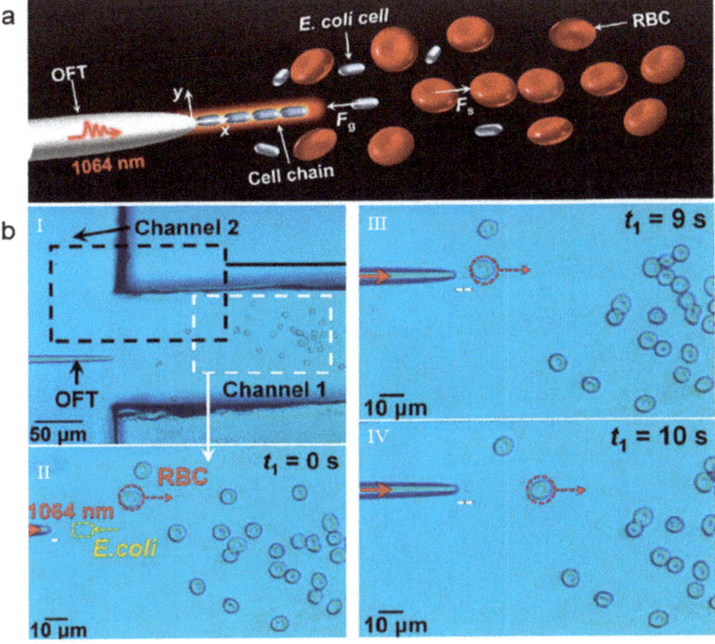

Figure 11. (**a**) Schematic of *E. coli* trapping and arrangement at the tip of the OFTs with red blood cells (RBCs) pushed away. (**b-I**) Laser OFF. In channel 1, the *E. coli* and RBCs are mixed together. (**b-II–IV**) Laser ON. The *E. coli* is attracted and the RBCs are released. Adapted with permission from Liu et al. [56].

6. Structured Optical Fiber for Trapping and Manipulation

The trapping and manipulation of different objects make a high requirement on structured OFTs, although single OFTs have been successfully used in many studies [57–59]. These structured OFTs will provide a new platform for optical manipulation, especially for the trapping of nanosized particles that need to overcome the diffraction limit. Structured OFTs will serve as a new candidate with much more powerful trapping and manipulation capabilities [4,60,61].

6.1. Internal Structured Optical Fiber Tweezers for Non-contact Trapping

6.1.1. Fiber-based Total Internal Refection Lens

There are many ways to achieve contactless capture, and the fiber-based total internal refection lens is one of them. In one case, a bundle of optical fiber was encapsulated in a quartz capillary and fixed with epoxy resin [62]. The end surface of optical fiber is properly processed to make the light propagation between the optical fiber and the surrounding medium in a total internal reflection form, so as to achieve the high numerical aperture focusing and realize the purpose of capture particles that are far away from the fiber end (Figure 12a). Moreover, the direction of light propagation is inclined to the z axis in this structure, and the exerted scattering force is therefore greatly weakened. A 3D optical trap that is constructed by the optical fiber with special internal structure can capture and manipulate objects in a non-contact and non-invasive manner. Different functions of this structure can be achieved and multiple potential wells can be implemented at different or same distances from the fiber. By changing the angle of the light, the trapped particles can be slightly moved when simultaneously collecting optical signals (Figure 12b).

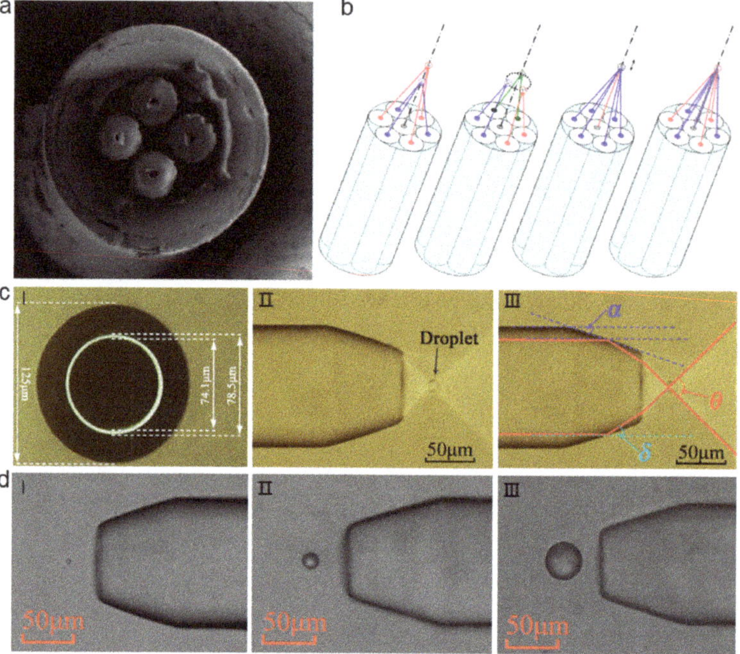

Figure 12. Fiber-based total internal refection lens for non-contact trapping. (**a**) Scanning electron microscope (SEM) image of the optical fiber probe. The gradient inclined hole is obtained by shaping the end of the fiber. (**b**) Different functions can be achieved by using optical fiber bundles. Adapted with permission from Liberale et al. [62]. (**c-I**) The profile view of the annular core fiber, the external diameter of annular fiber is 78.5 μm while the internal diameter is 74.1 μm, and the surrounding cladding diameter is 125 μm. (**c-II**) The lateral view of the annular core fiber. (**c-III**) Schematic diagram of the optical path. (**d**) Images of oil droplets captured by the annular core fiber optical tweezers. The diameters of the oil droplets are (**d-I**) 3.8 μm; (**d-II**) 16.4 μm; (**d-III**) 40.1 μm. Adapted with permission from Liu et al. [63].

In another method, an annular core fiber is used to make a new type of OFTs (Figure 12c-I) [63]. The end face of the annular core fiber is made into a special frustum cone shape, which can generate a strong optical force to capture droplets in a non-contact form. The light beam is completely reflected at the edge of the cone, and then deflected away from the end of the fiber. Therefore, objects can be captured and manipulated in a non-contact manner (Figure 12c-II,III). The special OFTs with a magnetic ring structure can capture a wide range of objects and droplets ranging from three microns to 40 microns in diameter. The shape of the trapped droplets can be perfectly maintained while using this non-contact method (Figure 12d).

6.1.2. Graded-Index Lens

Graded-index fiber can also achieve the non-contact capture and manipulation of objects. Developing adjustable OFTs can improve the operating ability and range of applications, which is conductive to its application of optical tweezers in more fields [44,64]. For example, the combination of OFTs and microfluidics can realize the adjustable working distance of optical tweezers [65,66].

Graded-index fiber taper has been proved to own a strong focus effect [66], and it enables optical trapping with adjustable working distances in liquid flow, combined with an optical microcavity (Figure 13). In this situation, the graded-index fiber and single-mode fiber are aligned in the capillary to form an air microcavity, the length of which can be adjusted by shifting the platform, and then directly affects the focusing ability. By adjusting the light focus or the balance between optical force and drag force, the adjustable OFTs are realized while improving the working range and operating ability (Figure 13b). The tip of graded-index fiber has lower transmission loss and the stronger transverse gradient force, which can enhance the operating range of OFTs. The graded-index OFTs are adjustable for non-contact trapping of particles with controllable working distance by changing the input power of light, flow rates, length of the micro-cavity, and other factors (Figure 13c).

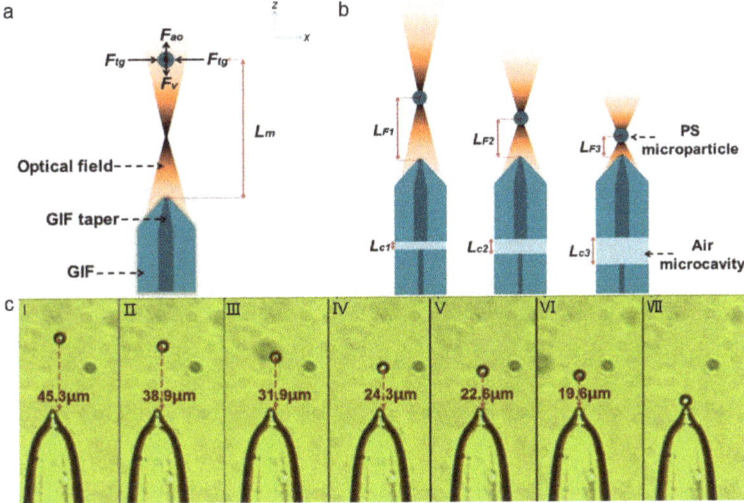

Figure 13. (**a**) Optofluidic tunable manipulation of microparticles by changing the laser power, flow rate and shape of the tapered fiber. (**b**) The operating range of OFTs can be adjusted by changing the length of the optical microcavity. The optical microcavity is the air cavity inside the fiber, as marked in aqua green and indicated by the dashed arrow. (**c**) Optofluidic tunable manipulation of microparticle with different working distance. Adapted with permission from Gong et al. [65].

6.2. Surface Structured OFTs for Nanomanipulation

The latest demand for nanotechnology is that it can precisely and noninvasively manipulate nanoscale objects [67]. However, trapping of objects with nanometer scale is of great challenge due to the diffraction limit and the strong Brownian motion [68,69]. Therefore, the nanomanipulation of OFTs has become the focus of researcher's attention [70]. Until now, different OFTs-based nanomanipulation techniques have been developed, such as plasmonic nanostructure-based, photonic nanojet-based, and connected fiber/nanoject combined forms.

6.2.1. Plasmonic Nanostructure-based Fiber Tweezers

A plasmonic nanostructured-based fiber tweezers can be used for nanomanipulation. In this case, a metal coating is applied to the tapered fiber and a bowtie plasmonic aperture is designed to capture and manipulate the nanoparticles (Figure 14a) [71]. The plasmonic nanostructures at the fiber end surface plays the key role in this technique, providing a powerful way of dealing with individual atoms that are attached to surfaces. The trapping mechanism of this optical fiber structure is based on the self-induction back-action, which can reduce the influence of the photothermal effect on the captured object [72]. The trapped specimens play an active role in the trapping mechanism, and the required field strength is several orders of magnitude weaker than traditional optical fiber tweezers. This technique works as a scanning optical microscope probe and nano-optical tweezers for nanomanipulation in three dimensions. The end of the probe is used to manipulate and detect objects, while the object is captured and detected while using optical fibers [70,73,74]. Using this plasmonic nanostructured fiber tweezers, sub-100-nm particles can be trapped and manipulated in three dimensions (Figure 14b).

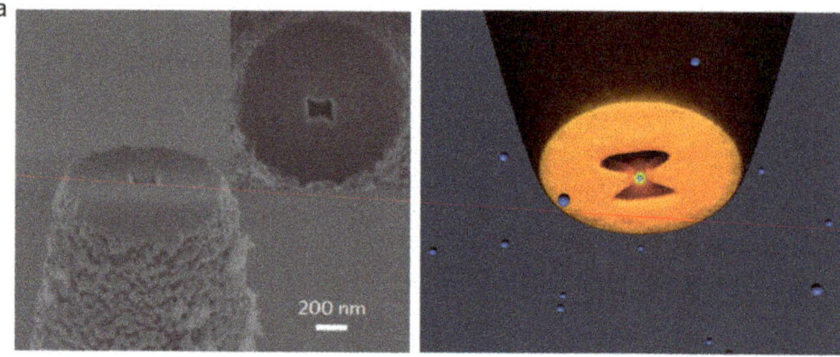

Figure 14. (**a**) SEM image of surface structured fiber with a plasmonic bowtie aperture. Adapted with permission from Pang et al. [71]. (**b**) Structure diagram for nanoparticle trapping on the bowtie nano-aperture structure.

6.2.2. Photonic Nanojet-based Fiber Tweezers

Currently, the selective capture of nanoscale targets and the simultaneous capture of multiple targets are facing huge challenges [67,75]. With high focusing characteristics, photonic nanojet can effectively reduce the critical size of the captured object [76,77]. Since being first reported by Chen et al. [78], photonic nanojets have been applied in many fields [79,80]. These photonic nanojets can be used for optical trapping and nanomanipulation when integrated into a fiber end. In this case, a microsphere that is positively charged is attached to the end of a tapered optical fiber (Figure 15a) [35]. Light is highly focused at the end of the sphere due to the photonic nanoject mechanism of the microsphere (Figure 15b). This light focusing can exert enhanced optical force on the nanoobject, and the shadow of the photonic nanojet surface forms a potential well for capturing nanoparticles in a non-contact way (Figure 15a). Therefore, this structure can be used for the trap of nanoparticles as well as biomolecules, such as DNA. Combining the three-dimensional optical manipulation with signal enhancement, this technique can also be used for the detection of nanoparticles and biomolecules.

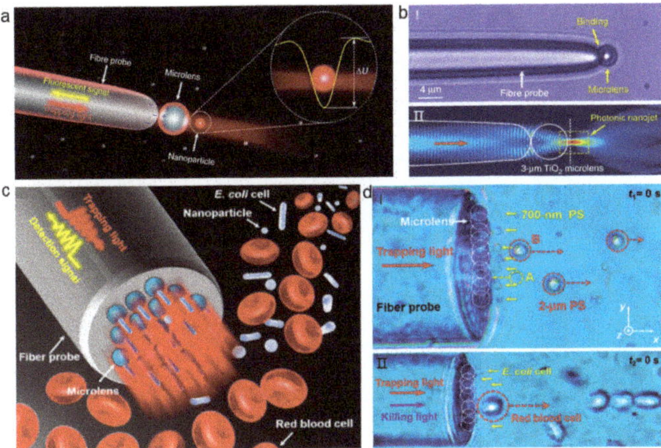

Figure 15. Photonic nanojet-based fiber tweezers. (**a**) Schematic of photonic nanojet-based trapping of nanoparticles. (**b**) Microscopic image (**I**) and simulation results of the photonic nanojet (**II**). Adapted with permission from Li et al. [81]. (**c**) Schematic of a parallel photonic nanojet array for selective trapping and manipulating of nanoparticles and cells in blood solution. (**d**) Images for photonic nanojet array trapping and separation of different particles and cells. Adapted with permission from Li et al. [82].

In addition, multiple dielectric microparticles can be arranged onto the surface of a fiber end to form photonic nanojet array [82]. The back of microparticles produces a highly focused beam because of the high refractive index and low light absorption of dielectric particles, which serves as photonic nanojet array, and can thus manipulate and detect nanoparticles and subwavelength cells. The photonic nanojet array assembled with microspheres can form tens to hundreds of nanowells and then manipulate and detect multiple nanoparticles or subwavelength cells simultaneously at low power (Figure 15c). The device owns a strong capture ability, while the optical power is far lower than the COTs, it therefore will not easily generate a photothermal effect. In addition, the optical signal that is transmitted through the front end of the optical fiber can also be detected through the same fiber in reflection. This can be used for the detection of different particles and cells. For different cells, the exerted force is different due to the different sizes. Scattering forces cause the large particles to move in the direction of light propagation, while the smaller bacteria cells can be trapped. This can be used for the selective trapping and separation of different cells (Figure 15d).

6.2.3. Connected Fiber/Nanojet Combined Fiber Tweezers

The connected fiber with combined nanojet at the surface can also be used for the manipulation of nanoparticles in addition to the solely surface modification of optical fiber end [83]. In this case, the single-mode fiber (SMF) and multi-mode fiber (MMF) are connected together to generate Bessel beam, which produces a narrow output laser, and a glass microsphere (GMS) is attached to the surface of the MMF to form a photonic nanojet (Figure 16a). At the tip of the optical fiber, the narrow laser beam is further focused into a nanoscale point by the glass spheres at high magnification (Figure 16b). This technique can thus be used for the trapping and manipulation of nanoparticles in a non-contact manner.

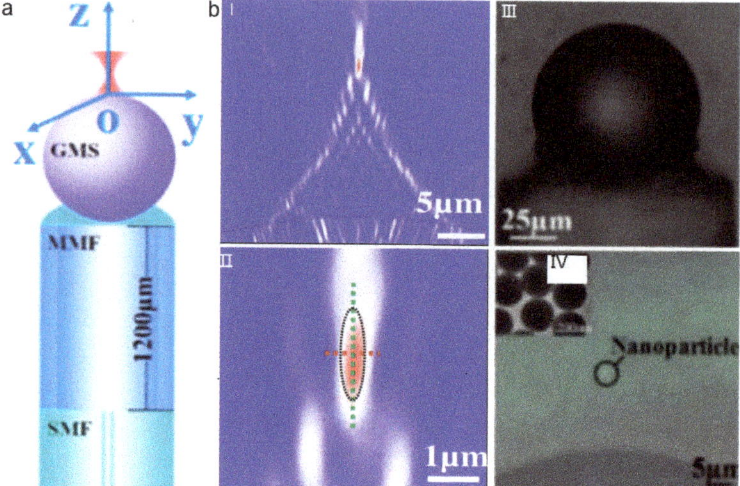

Figure 16. (**a**) Schematic of the fiber probe with a glass microsphere (GMS) attached. (**b**) (**I**),(**II**)Simulation results of light distribution; (**III**) Microscopic image of the probe. (**IV**) Optical trapping of nanoparticles with diameters of 200 nm. Adapted with permission from Tang et al. [83].

7. Subwavelength Optical Fiber Wire for Evanescent Fields-based Trapping and Delivery

Optical tweezers capture and manipulate particles with the optical force that is generated by a highly focused laser [84]. The diffraction limit and focusing depth of optical tweezers in space limit the applications of optical tweezers in the manipulation of particles and continuous delivery of particles in a long range. This problem can be solved when using a subwavelength optical fiber wire with evanescent fields at the surface, which can capture and manipulate particles in a long range [85]. Similar to the optical waveguides, optical forces that are generated by the evanescent field around the subwavelength fiber wire can also be used for the capture and delivery of particles [86–88].

For the subwavelength optical fiber wire, as shown in Figure 17a, near the optical fiber surface, evanescent fields decay exponentially away from the surface. At the fiber surface, the optical intensity is the maximum, and field gradient exists along the direction of the exponential decay, which is perpendicular to the fiber wire. With this field gradient, optical gradient force is generated perpendicular to the fiber surface, which can be used for particle trapping. Figure 17a also shows the calculated optical gradient force. Accordingly, the particle is stably trapped at the fiber surface by the optical gradient force. Along the fiber, there is an optical scattering force, which delivers the particles along the fiber surface in direction of light propagation [89]. Size dependent trapping and delivery of polystyrene spheres was achieved while using a 600 nm diameter fiber wire (Figure 17a). The optical force exerted on the polystyrene sphere increased with the increase of the diameter, and the delivery speed of the larger sphere was also higher than that of the smaller sphere. At low input laser power, the larger sphere was more likely to be captured on the surface of the fiber and then delivered along the propagation direction [90]. This subwavelength optical fiber wire can also be used for the manipulation of biological cells in addition to the manipulation and delivery of particles. For example, *E.coli* can be stably trapped on the surface of the fiber by optical gradient forces (Figure 17b) [91], and the trapped *E. coli* can further be delivered along the fiber, even in microfluidic environment.

In addition to the delivery in a straight trajectory along a straight fiber wire, particles can also be delivered along a bent optical fiber wire. For example, Li et al. demonstrated the delivery of nanoparticles along an arbitrary bent fiber wire (Figure 17c) [92]. The relationship between bending loss, bending radius, and center angle were studied. For a specific input power, the light capture and transmission had a corresponding minimum bending radius. The delivery of nanoparticles was first demonstrated by using 650 nm red light propagating along a 600-nm diameter optical fiber wire. The nanoparticles can be captured by the optical gradient force when the power of incident light was increased to 12 mW, and then delivered along the bent fiber in the direction of light propagation direction by the scattering force caused by evanescent field.

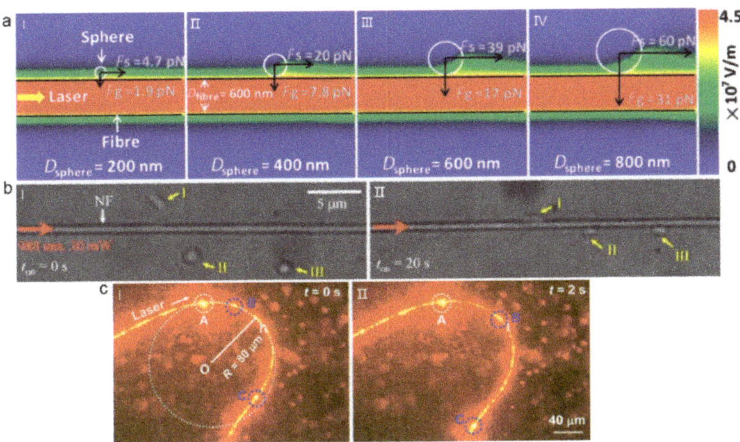

Figure 17. Evanescent field-based capture and delivery along optical fiber wire. (**a**) Simulation distribution of evanescent wave field around a fiber wire with different sized particles trapped. Adapted with permission from Xu et al. [90]. (**b**) Trapping of *E.coli* along an optical fiber wire. Adapted with permission from Xin et al. [91]. (**c**) Delivery of nanoparticles along bent optical fiber wire. Adapted with permission from Li et al. [92].

In addition to the delivery of particles in one direction along the fibre, bidirectional manipulation of particles can also be realized while using optical fiber wire [93,94]. In this case, two laser beams were transmitted in reverse direction into the fiber to realize bidirectional manipulation of particles. The magnitude and direction of the scattering force can be changed to control the delivery direction of the particles by changing the laser power of the input fiber (Figure 18) [82].

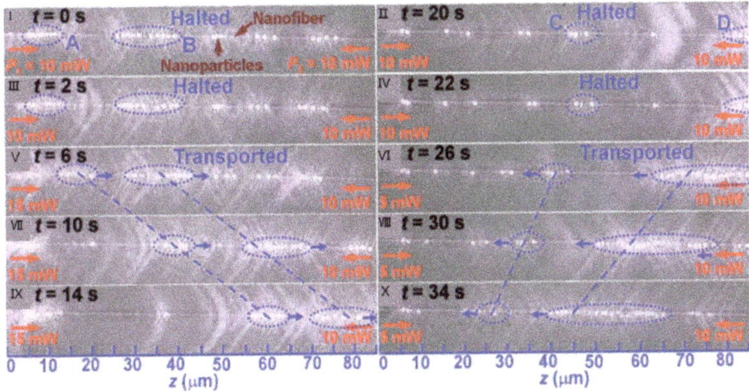

Figure 18. Bidirectional transportation and controllable positioning of nanoparticles. Adapted with permission from Lei et al. [93].

8. Optical Fiber-based Massive Photothermal Assembly

The photothermal effect due to light absorption can also be used for the trapping and manipulation of particles, in addition to the trapping and manipulation of particles by optical force using optical fiber tweezers. The light absorption of surrounding medium and particles can induce different photothermal effects. For the medium absorption, the main effect is the thermophoresis, where the temperature gradient that is induced by the light absorption can result in the particle migration responsive to temperature gradient by the Soret effect [95]. While for the particle absorption, photophoresis will induce the particle migration. The uneven photothermal phenomenon of light-irradiated particles can keep the high-absorptive particles away from the light-induced hot point [96,97], while the low-absorptive particles tend to the hot point, and the photothermal effect can, therefore, manipulate the large-scale particles [98].

For the photophoresis, one of the important parameters is the asymmetrical factor of heat distribution J, which can directly represent the heat flow [97]. The J factor is given by:

$$J = \frac{6\pi n_p \kappa_p}{n_f^2 R^3 \lambda} \cdot \int_0^R \int_0^\pi \frac{|E(r,\theta)|^2}{|E_0|^2} r^3 cos\theta sin\theta \cdot d\theta \cdot dr \tag{7}$$

where n_p is the refractive index of the particle, κ_p is the absorption rate of the particle, n_f is the refractive index of the solution, R is the radius of the particle, λ is the wavelength of incident light, $E(r,\theta)$ is the local electric field intensity within the particle, and E_0 is the electric field intensity of incident light. For particles in the liquid medium, the expression formula of the photophoretic velocity V_P is expressed, as follows [97]:

$$V_P = -\frac{\beta_T A r_0^2}{18\mu v_0 k_f} IJ \frac{ln3 + 4(ln3 - 1)\frac{L_S}{r_0}}{\left(\frac{k_p}{k_f} + 2\right)\left(1 + 2\frac{L_S}{R}\right)} \tag{8}$$

where β_T is the thermal expansion coefficient of solvent, A is Hamaker constant, r_0 is the radius of the solvent molecules, and μ is the viscous coefficient of liquid, v_0 is the solvent the characteristics of the molecular volume, k_f is the thermal conductivity solvent, I is light intensity, L_S is particle relative to the slip length of water, k_p is the thermal conductivity particles, and R is the particle radius. For a given solution and a particle of the same material, the photophoretic velocity V_P can be expressed, as follows (C is a constant, $C = -\frac{\beta_T A r_0^2}{18\mu v_0 k_f} \frac{ln3+4(ln3-1)\frac{L_S}{r_0}}{\left(\frac{k_p}{k_f}+2\right)}$) [99]:

$$V_P = \frac{CIJ}{\left(1 + 2\frac{L_S}{R}\right)} \tag{9}$$

In many situations, the photothermal effect is the combination of both the thermophoresis and the photophoresis. When compared with optical force, the force that is induced by photothermal effect is much stronger and, thus, can be used for massive particle trapping and assembly. For example, Lei et al. demonstrated the photothermal assembly of particles and *E. coli* while using a subwavelength diameter fiber (SDF) [100]. When the 1550 nm laser is launched into the fiber, the leaking light radiates the object and creates a negative photophoresis that pulls the object toward the fiber. The SDF functions as a linear "light source" and particles or other small objects around the SDF are exposed to the radiation. The energy of the leaking light will be absorbed by the particles and converted into a local heat distribution. The uneven heat distribution inside each sample causes the photothermal effect to drive the particles (Figure 19a). When suspended in medium, the particles will also move along the temperature gradient, usually from hot to cold. Particles move towards the strong region and gather in the center of the region, where the negative photophorsis force and the temperature gradient force are balanced, due to the combination effect of the photophoresis and thermophoresis. By moving the SDF, the assembled particles and bacteria can be migrated along the moving direction. Alternatively, a tapered optical fiber can also be used for the massive photothermal assembly of particles [98]. As shown in Figure 19b, particles can be assembled to the end of the fiber with a distance of several tens of microns due to the photothermal effect. By moving the fiber, the assembled particles can be migrated with an efficiency up to 95%.

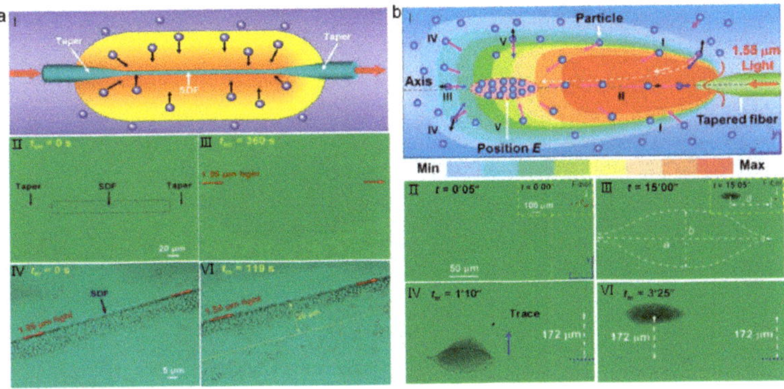

Figure 19. (**a**) Massive photothermal assembly and migration of particles and *E.coli* using subwavelength diameter fiber. (**b**) Assembly and migration of particles using a tapered optical fiber. Adapted with permission from Xin et al. [98].

9. Opto-Thermophoretic Fiber Tweezers

In addition to the optical force and photothermal effect, the electric fields can also contribute to the trapping and manipulation [101–104]. Kotnala et al. demonstrated an opto-thermophoretic fiber tweezer (OTFT) [105], which combined the photothermal effect and electric field effect into the fiber tweezers. These fiber tweezers are a new strategy for the low power manipulation of nanoscale objects. In OTFT, a layer of porous gold film was deposited on the fiber tip to form a thermoplasmonic fiber tip, which converts photons into phonons. The charged molecules were adsorbed onto particle surface to form a charged molecular layer, which forms into positive micelles, and Cl^- molecules work as counterions. With light irradiation, the photothermal effect induced thermophoresis results in the migration of particles. The positive micelles and Cl^- ions were separated, inducing an electric field between the fiber tip and the particles due to the different Soret coefficient, which is given by [105]:

$$E_T = \frac{k_B T \, \nabla T}{e} \frac{\sum_i Z_i n_i S_{Ti}}{\sum_i Z_i^2 n_i} \tag{10}$$

where i is the ionic species, k_B is the Boltzmann constant, T is the environmental temperature, ∇T is the temperature gradient, e is the elemental charge, Z_i is the chrage number, n_i is the concentration, and S_{Ti} is the Soret coefficient of i species. An electric field E_T pointing toward the fiber tip is generated, due to the higher molecular mass and larger Soret coefficient of the positive micelle. The resulting thermo-electric field results into the trapping of the positively charged particles (Figure 20a). OTFT uses opto-thermoelectric force to capture particles, which have the unique advantages as compared with normal OFTs: it can be used to capture wide range of nanoparticles with low power consumption (less than 1 mW). By increasing the power, the increased convective flow and thermophoretic force can be used for particle assembly and concentration (Figure 20b, power: 10 mW). By manipulating the fiber, the trapped nanoparticles can also be used for multifunction applications. For example, it can be used for the delivery of a single nanoparticle to a lipid vesicle membrane (Figure 20c).

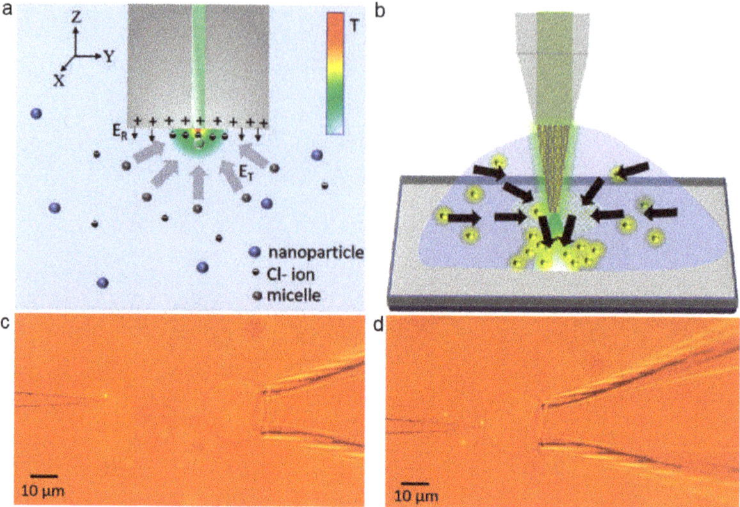

Figure 20. Opto-thermophoretic fiber tweezer (OTFT) for nanoparticle trapping and manipulation. Adapted with permission from Kotnala et al. [105]. (**a**) Schematic of OTFT, opto-thermoelectric force is used for nanoparticle trapping. (**b**) Schematic for nanoparticle assembly and concentration using OTFT. (**c**) and (**d**) OTFT serves as a nanopipette for single nanoparticle delivery.

10. Conclusions

In conclusion, this review presents recent development of optical fiber-based tweezers for optical trapping and manipulation. Different types of OFTs were discussed, including their corresponding principles and applications. Single OFTs enable the targeted trapping and flexible manipulation of particles and cells while using optical force, and enable single cell analysis. Evanescent field from subwavelength optical fiber wire enables the long-range manipulation and delivery of different targets. The photothermal effect of different fiber forms can be used for massive assembly and migration. The opto-thermalphoretic fiber tweezers are able to multifunctional trapping and manipulation of nanoparticles with low power consumption when combing the electric field. Fiber optical tweezers will provide many new possibilities for different applications in the near future due to the advantages of easy fabrication, compact size, and flexible manipulation.

Although great progress has been made in OFTs, there are still many open challenges, but also opportunities. First, the direct contact between the sample and fiber end surface might induce mechanical damage on the samples. Therefore, the non-contact and damage-free trapping technique based on OFTs is necessary to develop. Second, stable trapping and flexible manipulation of samples with sizes in the nanometer scale to overcome the diffraction limit while using OFTs is still a big challenge. In particular, stable trapping of single biomolecules is very difficult, but will be of great importance for single molecular analysis. Third, optical trapping and manipulation of cells and biological structures and the subsequent biosensing in vivo will be a new trend in the following few years. However, the OFTs may induce mechanical damage when inserting the fiber into living samples. Therefore, the construction of biocompatible OFTs is of great importance, and maintains great application potential. Living biophotonic waveguide via the assembly of living bacteria or cells in vivo will provide new possibilities for the formation of biocompatible optical fibers [53,54]. Trapping, manipulation, sensing, and diagnostics in vivo will be possible with such optical fibers.

Author Contributions: H.X. supervised the project; X.Z. and N.Z. wrote the original draft and prepared the figures; Y.S., H.X., and B.L. reviewed and edited the final manuscript. All authors have read and agreed to the published version of the manuscript.

Funding: This work was funded by the by the National Natural Science Foundation of China (61975065, 11904132), Guangdong Basic and Applied Basic Research Foundation (2019B151502035), the Fundamental Research Funds for the Central Universities (21619323).

Conflicts of Interest: The authors declare no conflict of interest.

References

1. Ashkin, A. Acceleration and trapping of particles by radiation pressure. *Phys. Rev. Lett.* **1970**, *24*, 156. [CrossRef]
2. Ashkin, A.; Dziedzic, J.M.; Bjorkholm, J.; Chu, S. Observation of a single-beam gradient force optical trap for dielectric particles. *Opt. Lett.* **1986**, *11*, 288–290. [CrossRef] [PubMed]
3. Ashkin, A.; Dziedzic, J.M. Optical trapping and manipulation of viruses and bacteria. *Science* **1987**, *235*, 1517–1520. [CrossRef] [PubMed]
4. Baker, J.E.; Badman, R.P.; Wang, M.D. Nanophotonic trapping: Precise manipulation and measurement of biomolecular arrays. *Wiley Interdiscip. Rev. Nanomed. Nanobiotechnol.* **2018**, *10*, e1477. [CrossRef] [PubMed]
5. Choudhary, D.; Mossa, A.; Jadhav, M.; Cecconi, C. Bio-molecular applications of recent developments in optical tweezers. *Biomolecules* **2019**, *9*, 23. [CrossRef] [PubMed]
6. Dufresne, E.R.; Grier, D.G. Optical tweezer arrays and optical substrates created with diffractive optics. *Rev. Sci. Instrum.* **1998**, *69*, 1974–1977. [CrossRef]
7. Korda, P.; Spalding, G.C.; Dufresne, E.R.; Grier, D.G. Nanofabrication with holographic optical tweezers. *Rev. Sci. Instrum.* **2002**, *73*, 1956–1957. [CrossRef]
8. Plewa, J.; Tanner, E.; Mueth, D.M.; Grier, D.G. Processing carbon nanotubes with holographic optical tweezers. *Opt. Express* **2004**, *12*, 1978–1981. [CrossRef]
9. Chapin, S.C.; Germain, V.; Dufresne, E.R. Automated trapping, assembly, and sorting with holographic optical tweezers. *Opt. Express* **2006**, *14*, 13095–13100. [CrossRef]
10. Sun, B.; Roichman, Y.; Grier, D.G. Theory of holographic optical trapping. *Opt. Express* **2008**, *16*, 15765–15776. [CrossRef]
11. Quidant, R.; Girard, C. Surface-plasmon-based optical manipulation. *Laser Photonics Rev.* **2008**, *2*, 47–57. [CrossRef]
12. Righini, M.; Volpe, G.; Girard, C.; Petrov, D.; Quidant, R. Surface plasmon optical tweezers: Tunable optical manipulation in the femtonewton range. *Phys. Rev. Lett.* **2008**, *100*, 186804. [CrossRef] [PubMed]
13. Zhang, W.; Huang, L.; Santschi, C.; Martin, O.J. Trapping and sensing 10 nm metal nanoparticles using plasmonic dipole antennas. *Nano Lett.* **2010**, *10*, 1006–1011. [CrossRef] [PubMed]
14. Pang, Y.; Gordon, R. Optical trapping of a single protein. *Nano Lett.* **2011**, *12*, 402–406. [CrossRef]
15. Fuh, M.-R.S.; Burgess, L.W.; Hirschfeld, T.; Christian, G.D.; Wang, F. Single fibre optic fluorescence pH probe. *Analyst* **1987**, *112*, 1159–1163. [CrossRef]
16. Hu, Z.; Wang, J.; Liang, J. Manipulation and arrangement of biological and dielectric particles by a lensed fiber probe. *Opt. Express* **2004**, *12*, 4123–4128. [CrossRef]
17. Ribeiro, R.S.R.; Soppera, O.; Oliva, A.G.; Guerreiro, A.; Jorge, P.A. New trends on optical fiber tweezers. *J. Lightwave Technol.* **2015**, *33*, 3394–3405. [CrossRef]
18. Constable, A.; Kim, J.; Mervis, J.; Zarinetchi, F.; Prentiss, M. Demonstration of a fiber-optical light-force trap. *Opt. Lett.* **1993**, *18*, 1867–1869. [CrossRef]
19. Jensen-McMullin, C.; Lee, H.P.; Lyons, E.R. Demonstration of trapping, motion control, sensing and fluorescence detection of polystyrene beads in a multi-fiber optical trap. *Opt. Express* **2005**, *13*, 2634–2642. [CrossRef]
20. Taguchi, K.; Ueno, H.; Hiramatsu, T.; Ikeda, M. Optical trapping of dielectric particle and biological cell using optical fibre. *Electron. Lett.* **1997**, *33*, 413–414. [CrossRef]
21. Lyons, E.; Sonek, G. Confinement and bistability in a tapered hemispherically lensed optical fiber trap. *Appl. Phys. Lett.* **1995**, *66*, 1584–1586. [CrossRef]
22. Taguchi, K.; Ueno, H.; Ikeda, M. Rotational manipulation of a yeast cell using optical fibres. *Electron. Lett.* **1997**, *33*, 1249–1250. [CrossRef]
23. Taguchi, K.; Atsuta, K.; Nakata, T.; Ikeda, M. Levitation of a microscopic object using plural optical fibers. *Opt. Commun.* **2000**, *176*, 43–47. [CrossRef]

24. Zharov, V.P.; Kurten, R.C.; Bauman, J. Photothermal tweezers. *Proc. SPIE Int. Soc. Opt. Eng.* **2003**, *4960*, 134–141.
25. Neuman, K.C.; Block, S.M. Optical trapping. *Rev. Sci. Instrum.* **2004**, *75*, 2787–2809. [CrossRef]
26. Gherardi, D.M.; Carruthers, A.E.; Čižmár, T.; Wright, E.M.; Dholakia, K. A dual beam photonic crystal fiber trap for microscopic particles. *Appl. Phys. Lett.* **2008**, *93*, 041110. [CrossRef]
27. Jess, P.; Garcés-Chávez, V.; Smith, D.; Mazilu, M.; Paterson, L.; Riches, A.; Herrington, C.; Sibbett, W.; Dholakia, K. Dual beam fibre trap for Raman microspectroscopy of single cells. *Opt. Express* **2006**, *14*, 5779–5791. [CrossRef]
28. Decombe, J.-B.; Huant, S.; Fick, J. Single and dual fiber nano-tip optical tweezers: Trapping and analysis. *Opt. Express* **2013**, *21*, 30521–30531. [CrossRef]
29. Black, B.J.; Mohanty, S.K. Fiber-optic spanner. *Opt. Lett.* **2012**, *37*, 5030–5032. [CrossRef]
30. Kreysing, M.; Ott, D.; Schmidberger, M.J.; Otto, O.; Schürmann, M.; Martín-Badosa, E.; Whyte, G.; Guck, J. Dynamic operation of optical fibres beyond the single-mode regime facilitates the orientation of biological cells. *Nat. Commun.* **2014**, *5*, 5481. [CrossRef]
31. Guck, J.; Ananthakrishnan, R.; Mahmood, H.; Moon, T.J.; Cunningham, C.C.; Käs, J. The optical stretcher: A novel laser tool to micromanipulate cells. *Biophys. J.* **2001**, *81*, 767–784. [CrossRef]
32. Franze, K.; Grosche, J.; Skatchkov, S.N.; Schinkinger, S.; Foja, C.; Schild, D.; Uckermann, O.; Travis, K.; Reichenbach, A.; Guck, J. Müller cells are living optical fibers in the vertebrate retina. *Proc. Natl. Acad. Sci. USA* **2007**, *104*, 8287–8292. [CrossRef] [PubMed]
33. Rancourt-Grenier, S.; Wei, M.-T.; Bai, J.-J.; Chiou, A.; Bareil, P.P.; Duval, P.-L.; Sheng, Y. Dynamic deformation of red blood cell in dual-trap optical tweezers. *Opt. Express* **2010**, *18*, 10462–10472. [CrossRef] [PubMed]
34. Xin, H.; Li, B. Optical orientation and shifting of a single multiwalled carbon nanotube. *Light Sci. Appl.* **2014**, *3*, e205. [CrossRef]
35. Xu, X.; Cheng, C.; Zhang, Y.; Lei, H.; Li, B. Dual focused coherent beams for three-dimensional optical trapping and continuous rotation of metallic nanostructures. *Sci. Rep.* **2016**, *6*, 29449. [CrossRef]
36. Liu, Y.; Yu, M. Investigation of inclined dual-fiber optical tweezers for 3D manipulation and force sensing. *Opt. Express* **2009**, *17*, 13624–13638. [CrossRef]
37. Asadollahbaik, A.; Thiele, S.; Weber, K.; Kumar, A.; Drozella, J.; Sterl, F.; Herkommer, A.M.; Giessen, H.; Fick, J. Highly Efficient Dual-Fiber Optical Trapping with 3D Printed Diffractive Fresnel Lenses. *ACS Photonics* **2019**. [CrossRef]
38. Xin, H.; Xu, R.; Li, B. Optical trapping, driving, and arrangement of particles using a tapered fibre probe. *Sci. Rep.* **2012**, *2*, 818. [CrossRef]
39. Liu, Z.; Guo, C.; Yang, J.; Yuan, L. Tapered fiber optical tweezers for microscopic particle trapping: Fabrication and application. *Opt. Express* **2006**, *14*, 12510–12516. [CrossRef]
40. Grier, D.G. A revolution in optical manipulation. *Nature* **2003**, *424*, 810. [CrossRef]
41. Xin, H.; Liu, Q.; Li, B. Non-contact fiber-optical trapping of motile bacteria: Dynamics observation and energy estimation. *Sci. Rep.* **2014**, *4*, 6576. [CrossRef] [PubMed]
42. Yuan, L.; Liu, Z.; Yang, J.; Guan, C. Twin-core fiber optical tweezers. *Opt. Express* **2008**, *16*, 4559–4566. [CrossRef] [PubMed]
43. Mohanty, S.K.; Mohanty, K.; Berns, M.W. Manipulation of mammalian cells using a single-fiber optical microbeam. *J. Biomed. Opt.* **2008**, *13*, 054049. [CrossRef] [PubMed]
44. Xin, H.; Li, Y.; Li, L.; Xu, R.; Li, B. Optofluidic manipulation of Escherichia coli in a microfluidic channel using an abruptly tapered optical fiber. *Appl. Phys. Lett.* **2013**, *103*, 033703. [CrossRef]
45. Mestres, P.; Berthelot, J.; Spasenović, M.; Gieseler, J.; Novotny, L.; Quidant, R. Cooling and manipulation of a levitated nanoparticle with an optical fiber trap. *Appl. Phys. Lett.* **2015**, *107*, 151102. [CrossRef]
46. Xin, H.; Li, Y.; Xu, D.; Zhang, Y.; Chen, C.H.; Li, B. Single Upconversion Nanoparticle–Bacterium Cotrapping for Single-Bacterium Labeling and Analysis. *Small* **2017**, *13*, 1603418. [CrossRef] [PubMed]
47. Derby, B. Printing and prototyping of tissues and scaffolds. *Science* **2012**, *338*, 921–926. [CrossRef]
48. Saltzman, W.M.; Olbricht, W.L. Building drug delivery into tissue engineering design. *Nat. Rev. Drug Discov.* **2002**, *1*, 177. [CrossRef]
49. Chen, Z.; Li, Y.; Liu, W.; Zhang, D.; Zhao, Y.; Yuan, B.; Jiang, X. Patterning mammalian cells for modeling three types of naturally occurring cell–cell interactions. *Angew. Chem. Int. Ed.* **2009**, *48*, 8303–8305. [CrossRef]

50. Tam, J.M.; Biran, I.; Walt, D.R. An imaging fiber-based optical tweezer array for microparticle array assembly. *Appl. Phys. Lett.* **2004**, *84*, 4289–4291. [CrossRef]

51. Xin, H.; Xu, R.; Li, B. Optical formation and manipulation of particle and cell patterns using a tapered optical fiber. *Laser Photonics Rev.* **2013**, *7*, 801–809. [CrossRef]

52. Li, Y.; Xin, H.; Liu, X.; Li, B. Non-contact intracellular binding of chloroplasts in vivo. *Sci. Rep.* **2015**, *5*, 10925. [CrossRef] [PubMed]

53. Xin, H.; Li, Y.; Liu, X.; Li, B. Escherichia coli-based biophotonic waveguides. *Nano Lett.* **2013**, *13*, 3408–3413. [CrossRef] [PubMed]

54. Li, Y.; Xin, H.; Zhang, Y.; Lei, H.; Zhang, T.; Ye, H.; Saenz, J.J.; Qiu, C.-W.; Li, B. Living nanospear for near-field optical probing. *ACS Nano* **2018**, *12*, 10703–10711. [CrossRef]

55. Li, Y.; Liu, X.; Yang, X.; Lei, H.; Zhang, Y.; Li, B. Enhancing upconversion fluorescence with a natural bio-microlens. *ACS Nano* **2017**, *11*, 10672–10680. [CrossRef]

56. Liu, S.; Li, Z.; Weng, Z.; Li, Y.; Shui, L.; Jiao, Z.; Chen, Y.; Luo, A.; Xing, X.; He, S. Miniaturized optical fiber tweezers for cell separation by optical force. *Opt. Lett.* **2019**, *44*, 1868–1871. [CrossRef]

57. Svoboda, K.; Mitra, P.P.; Block, S.M. Fluctuation analysis of motor protein movement and single enzyme kinetics. *Proc. Natl. Acad. Sci. USA* **1994**, *91*, 11782–11786. [CrossRef]

58. Mitchem, L.; Reid, J.P. Optical manipulation and characterisation of aerosol particles using a single-beam gradient force optical trap. *Chem. Soc. Rev.* **2008**, *37*, 756–769. [CrossRef]

59. Dholakia, K.; Čižmár, T. Shaping the future of manipulation. *Nat. Photonics* **2011**, *5*, 335. [CrossRef]

60. Erickson, D.; Serey, X.; Chen, Y.-F.; Mandal, S. Nanomanipulation using near field photonics. *Lab Chip* **2011**, *11*, 995–1009. [CrossRef]

61. Lin, S.; Zhu, W.; Jin, Y.; Crozier, K.B. Surface-enhanced raman scattering with ag nanoparticles optically trapped by a photonic crystal cavity. *Nano Lett.* **2013**, *13*, 559–563. [CrossRef] [PubMed]

62. Liberale, C.; Minzioni, P.; Bragheri, F.; De Angelis, F.; Di Fabrizio, E.; Cristiani, I. Miniaturized all-fibre probe for three-dimensional optical trapping and manipulation. *Nat. Photonics* **2007**, *1*, 723. [CrossRef]

63. Liu, Z.; Chen, Y.; Zhao, L.; Zhang, Y.; Wei, Y.; Zhu, Z.; Yang, J.; Yuan, L. Single fiber optical trapping of a liquid droplet and its application in microresonator. *Opt. Commun.* **2016**, *381*, 371–376. [CrossRef]

64. Hester, B.; Campbell, G.K.; López-Mariscal, C.; Filgueira, C.L.; Huschka, R.; Halas, N.J.; Helmerson, K. Tunable optical tweezers for wavelength-dependent measurements. *Rev. Sci. Instrum.* **2012**, *83*, 043114. [CrossRef]

65. Gong, Y.; Zhang, C.; Liu, Q.-F.; Wu, Y.; Wu, H.; Rao, Y.; Peng, G.-D. Optofluidic tunable manipulation of microparticles by integrating graded-index fiber taper with a microcavity. *Opt. Express* **2015**, *23*, 3762–3769. [CrossRef]

66. Gong, Y.; Huang, W.; Liu, Q.-F.; Wu, Y.; Rao, Y.; Peng, G.-D.; Lang, J.; Zhang, K. Graded-index optical fiber tweezers with long manipulation length. *Opt. Express* **2014**, *22*, 25267–25276. [CrossRef]

67. Maragò, O.M.; Jones, P.H.; Gucciardi, P.G.; Volpe, G.; Ferrari, A.C. Optical trapping and manipulation of nanostructures. *Nat. Nanotechnol.* **2013**, *8*, 807. [CrossRef]

68. Yoo, D.; Gurunatha, K.L.; Choi, H.-K.; Mohr, D.A.; Ertsgaard, C.T.; Gordon, R.; Oh, S.-H. Low-power optical trapping of nanoparticles and proteins with resonant coaxial nanoaperture using 10 nm gap. *Nano Lett.* **2018**, *18*, 3637–3642. [CrossRef]

69. Pang, Y.; Gordon, R. Optical trapping of 12 nm dielectric spheres using double-nanoholes in a gold film. *Nano Lett.* **2011**, *11*, 3763–3767. [CrossRef]

70. Martin, O.J.; Girard, C. Controlling and tuning strong optical field gradients at a local probe microscope tip apex. *Appl. Phys. Lett.* **1997**, *70*, 705–707. [CrossRef]

71. Berthelot, J.; Aćimović, S.; Juan, M.; Kreuzer, M.; Renger, J.; Quidant, R. Three-dimensional manipulation with scanning near-field optical nanotweezers. *Nat. Nanotechnol.* **2014**, *9*, 295–299. [CrossRef] [PubMed]

72. Juan, M.L.; Gordon, R.; Pang, Y.; Eftekhari, F.; Quidant, R. Self-induced back-action optical trapping of dielectric nanoparticles. *Nat. Phys.* **2009**, *5*, 915. [CrossRef]

73. Chen, C.; Juan, M.L.; Li, Y.; Maes, G.; Borghs, G.; Van Dorpe, P.; Quidant, R. Enhanced optical trapping and arrangement of nano-objects in a plasmonic nanocavity. *Nano Lett.* **2011**, *12*, 125–132. [CrossRef] [PubMed]

74. Descharmes, N.; Dharanipathy, U.P.; Diao, Z.; Tonin, M.; Houdré, R. Observation of backaction and self-induced trapping in a planar hollow photonic crystal cavity. *Phys. Rev. Lett.* **2013**, *110*, 123601. [CrossRef] [PubMed]

75. Kang, P.; Serey, X.; Chen, Y.-F.; Erickson, D. Angular orientation of nanorods using nanophotonic tweezers. *Nano Lett.* **2012**, *12*, 6400–6407. [CrossRef] [PubMed]

76. Li, X.; Chen, Z.; Taflove, A.; Backman, V. Optical analysis of nanoparticles via enhanced backscattering facilitated by 3-d photonic nanojets. *Opt. Express* **2005**, *13*, 526–533. [CrossRef] [PubMed]

77. Ferrand, P.; Wenger, J.; Devilez, A.; Pianta, M.; Stout, B.; Bonod, N.; Popov, E.; Rigneault, H. Direct imaging of photonic nanojets. *Opt. Express* **2008**, *16*, 6930–6940. [CrossRef]

78. Chen, Z.; Taflove, A.; Backman, V. Photonic nanojet enhancement of backscattering of light by nanoparticles: A potential novel visible-light ultramicroscopy technique. *Opt. Express* **2004**, *12*, 1214–1220. [CrossRef]

79. Wang, Z.; Guo, W.; Li, L.; Luk'yanchuk, B.; Khan, A.; Liu, Z.; Chen, Z.; Hong, M. Optical virtual imaging at 50 nm lateral resolution with a white-light nanoscope. *Nat. Commun.* **2011**, *2*, 218. [CrossRef]

80. Mcleod, E.; Arnold, C.B. Subwavelength direct-write nanopatterning using optically trapped microspheres. *Nat. Nanotechnol.* **2008**, *3*, 413. [CrossRef]

81. Li, Y.-C.; Xin, H.-B.; Lei, H.-X.; Liu, L.-L.; Li, Y.-Z.; Zhang, Y.; Li, B.-J. Manipulation and detection of single nanoparticles and biomolecules by a photonic nanojet. *Light Sci. Appl.* **2016**, *5*, e16176. [CrossRef] [PubMed]

82. Li, Y.; Xin, H.; Liu, X.; Zhang, Y.; Lei, H.; Li, B. Trapping and detection of nanoparticles and cells using a parallel photonic nanojet array. *ACS Nano* **2016**, *10*, 5800–5808. [CrossRef] [PubMed]

83. Tang, X.; Zhang, Y.; Su, W.; Zhang, Y.; Liu, Z.; Yang, X.; Zhang, J.; Yang, J.; Yuan, L. Super-low-power optical trapping of a single nanoparticle. *Opt. Lett.* **2019**, *44*, 5165–5168. [CrossRef]

84. Jiang, H.-R.; Wada, H.; Yoshinaga, N.; Sano, M. Manipulation of colloids by a nonequilibrium depletion force in a temperature gradient. *Phys. Rev. Lett.* **2009**, *102*, 208301. [CrossRef] [PubMed]

85. Brambilla, G.; Murugan, G.S.; Wilkinson, J.; Richardson, D. Optical manipulation of microspheres along a subwavelength optical wire. *Opt. Lett.* **2007**, *32*, 3041–3043. [CrossRef]

86. Xin, H.; Li, B. Multi-destination release of nanoparticles using an optical nanofiber assisted by a barrier. *AIP Adv.* **2012**, *2*, 012166. [CrossRef]

87. Yang, A.H.; Moore, S.D.; Schmidt, B.S.; Klug, M.; Lipson, M.; Erickson, D. Optical manipulation of nanoparticles and biomolecules in sub-wavelength slot waveguides. *Nature* **2009**, *457*, 71. [CrossRef]

88. Schmidt, B.S.; Yang, A.H.; Erickson, D.; Lipson, M. Optofluidic trapping and transport on solid core waveguides within a microfluidic device. *Opt. Express* **2007**, *15*, 14322–14334. [CrossRef]

89. Xin, H.; Li, B. Targeted delivery and controllable release of nanoparticles using a defect-decorated optical nanofiber. *Opt. Express* **2011**, *19*, 13285–13290. [CrossRef]

90. Xu, L.; Li, Y.; Li, B. Size-dependent trapping and delivery of submicro-spheres using a submicrofibre. *New J. Phys.* **2012**, *14*, 033020. [CrossRef]

91. Xin, H.; Cheng, C.; Li, B. Trapping and delivery of escherichia coli in a microfluidic channel using an optical nanofiber. *Nanoscale* **2013**, *5*, 6720–6724. [CrossRef] [PubMed]

92. Li, Y.; Xu, L.; Li, B. Optical delivery of nanospheres using arbitrary bending nanofibers. *J. Nanopart. Res.* **2012**, *14*, 799. [CrossRef]

93. Lei, H.; Xu, C.; Zhang, Y.; Li, B. Bidirectional optical transportation and controllable positioning of nanoparticles using an optical nanofiber. *Nanoscale* **2012**, *4*, 6707–6709. [CrossRef] [PubMed]

94. Zhang, Y.; Li, B. Particle sorting using a subwavelength optical fiber. *Laser Photonics Rev.* **2013**, *7*, 289–296. [CrossRef]

95. Vigolo, D.; Rusconi, R.; Stone, H.A.; Piazza, R. Thermophoresis: Microfluidics characterization and separation. *Soft Matter* **2010**, *6*, 3489–3493. [CrossRef]

96. Jovanovic, O. Photophoresis—Light induced motion of particles suspended in gas. *J. Quant. Spectrosc. Radiat. Transf.* **2009**, *110*, 889–901. [CrossRef]

97. Soong, C.; Li, W.; Liu, C.; Tzeng, P. Theoretical analysis for photophoresis of a microscale hydrophobic particle in liquids. *Opt. Express* **2010**, *18*, 2168–2182. [CrossRef]

98. Xin, H.; Li, X.; Li, B. Massive photothermal trapping and migration of particles by a tapered optical fiber. *Opt. Express* **2011**, *19*, 17065–17074. [CrossRef]

99. Xin, H.; Bao, D.; Zhong, F.; Li, B. Photophoretic separation of particles using two tapered optical fibers. *Laser Phys. Lett.* **2013**, *10*, 036004. [CrossRef]

100. Lei, H.; Zhang, Y.; Li, X.; Li, B. Photophoretic assembly and migration of dielectric particles and escherichia coli in liquids using a subwavelength diameter optical fiber. *Lab Chip* **2011**, *11*, 2241–2246. [CrossRef]

101. Wu, M.C. Optoelectronic tweezers. *Nat. Photonics* **2011**, *5*, 322. [CrossRef]

102. Braun, M.; Würger, A.; Cichos, F. Trapping of single nano-objects in dynamic temperature fields. *Phys. Chem. Chem. Phys.* **2014**, *16*, 15207–15213. [CrossRef] [PubMed]
103. Braun, M.; Bregulla, A.P.; Günther, K.; Mertig, M.; Cichos, F. Single molecules trapped by dynamic inhomogeneous temperature fields. *Nano Lett.* **2015**, *15*, 5499–5505. [CrossRef] [PubMed]
104. Lin, L.; Hill, E.H.; Peng, X.; Zheng, Y. Optothermal manipulations of colloidal particles and living cells. *Acc. Chem. Res.* **2018**, *51*, 1465–1474. [CrossRef] [PubMed]
105. Kotnala, A.; Zheng, Y. Opto-thermophoretic fiber tweezers. *Nanophotonics* **2019**, *8*, 475–485. [CrossRef]

 *micromachines*

Editorial

Editorial for the Special Issue on Optical Trapping and Manipulation: From Fundamentals to Applications

Daniel R. Burnham [1,*] and Philip H. Jones [2,*]

1   The Francis Crick Institute, 1 Midland Road, London NW1 1AT, UK
2   Department of Physics & Astronomy, University College London, Gower Street, London WC1E 6BT, UK
*   Correspondence: Daniel.Burnham@crick.ac.uk (D.R.B.); philip.jones@ucl.ac.uk (P.H.J.)

Received: 27 March 2020; Accepted: 14 April 2020; Published: 15 April 2020

This Special Issue of *Micromachines* is devoted to optical trapping, and the enormous range of uses the method has found in the decades since its first demonstration. The papers published here include both novel research articles and in-depth reviews, innovative optical trapping schemes and notable new applications. This diversity of the research field is reflected in the breadth of papers contained in this Special Issue.

Two papers are concerned with the investigation of mechanical interactions of red blood cells (RBCs). Zhu et al. have investigated the disaggregation of RBCs by using optical tweezers to quantify the force required to separate aggregated cells, showing that a short irradiation from a pulsed helium-neon laser may reduce the aggregation force [1]. Avsievich et al. report on the effect of polymeric nanocapsules—a model for drug delivery vehicles—on the aggregation force of RBCs, finding no change and no cytotoxicity as a result of the nanocapsule treatment [2].

Vivek et al. have used optical tweezers to study biomembranes, by inducing the controlled fusion of giant unilamellar vesicles (GUVs) [3]. The manipulation of membrane components of such 'artificial cells' provides great insight into the fusion process and morphological transitions that occur during fusion, and the authors exploit the spatial and temporal control provided by optical trapping to observe the fine details of fusion events.

Optical tweezers also find application in single-molecule biophysics, such as the work reported by Kretzer et al. [4] By combining optical tweezers into a microfluidic device, the authors were able to measure the concentration-dependent effects on the mechanics of single molecule ligand binding.

A novel optical trapping scheme utilizing a micro-ring resonator is reported by Ho et al., who used the evanescent field of the device to trap microscopic particles [5]. The system uses the resonant enhancement of the field in the resonator, together with a closed-loop system that renders the trapping insensitive to changes in the environment.

In their communication, Shishkin et al. reported on the development of microscopic optomechanical tools for use with optical tweezers [6]. They demonstrate the fabrication of a clamping fork that can be used to hold cells and is capable or arbitrary three-dimensional manipulation and also rotations. Such microtools are advantageous for enhancing the range of manipulations that may be applied to trapped objects.

Microscopic optically trapped and actuated machines are considered in depth in the review article by Andrew et al. [7]. The authors give a motivation for the use of such microrobots for avoiding optical damage to delicate biological specimens, and present the evolution of a number of different designs for a range of applications.

Finally, Zhao et al. have provided an in-depth review of an alternative optical trapping scheme, namely optical fibre tweezers [8]. They have provided a comprehensive overview of a number of different optical trapping configurations that use optical fibres, including dual fibre, single fibre and structured fibre traps; the benefits such schemes may have over conventional optical tweezers, and the range of applications they have found.

We would like to thank all the authors who have contributed their work to this Special Issue, and all the referees who have dedicated their time to the rigorous reviewing process.

In closing this editorial, we would like to remind readers that a second volume of *Optical Trapping and Manipulation: from Fundamentals to Applications* is accepting contributions until 31 January 2021, for publication later that year.

Conflicts of Interest: The authors declare no conflict of interest.

References

1. Zhu, R.; Avsievich, T.; Bykov, A.; Popov, A.; Meglinski, I. Influence of Pulsed He–Ne Laser Irradiation on the Red Blood Cell Interaction Studied by Optical Tweezers. *Micromachines* **2019**, *10*, 853. [CrossRef] [PubMed]
2. Avsievich, T.; Tarakanchikova, Y.; Zhu, R.; Popov, A.; Bykov, A.; Skovorodkin, I.; Vainio, S.; Meglinski, I. Impact of Nanocapsules on Red Blood Cells Interplay Jointly Assessed by Optical Tweezers and Microscopy. *Micromachines* **2020**, *11*, 19. [CrossRef] [PubMed]
3. Vivek, A.; Bolognesi, G.; Elani, Y. Fusing Artificial Cell Components and Lipid Domains Using Optical Traps: A Tool to Modulate Membrane Composition and Phase Behavior. *Micromachines* **2020**, *11*, 388. [CrossRef] [PubMed]
4. Kretzer, B.; Kiss, B.; Tordai, H.; Csík, G.; Herényi, L.; Kellermayer, M. Single-Molecule Mechanics in Ligand Concentration Gradient. *Micromachines* **2020**, *11*, 212. [CrossRef] [PubMed]
5. Ho, V.W.L.; Chang, Y.; Liu, Y.; Zhang, C.; Li, Y.; Davidson, R.R.; Little, B.E.; Wang, G.; Chu, S.T. Optical Trapping and Manipulating with a Silica Microring Resonator in a Self-Locked Scheme. *Micromachines* **2020**, *11*, 202. [CrossRef] [PubMed]
6. Shishkin, I.; Markovich, H.; Roichman, Y.; Ginzburg, P. Auxiliary Optomechanical Tools for 3D Cell Manipulation. *Micromachines* **2020**, *11*, 90. [CrossRef] [PubMed]
7. Andrew, P.-K.; Williams, M.A.K.; Avci, E. Optical Micromachines for Biological Studies. *Micromachines* **2020**, *11*, 192. [CrossRef] [PubMed]
8. Zhao, X.; Zhao, N.; Shi, Y.; Xin, H.; Li, B. Optical Fiber Tweezers: A Versatile Tool for Optical Trapping and Manipulation. *Micromachines* **2020**, *11*, 114. [CrossRef] [PubMed]

MDPI
St. Alban-Anlage 66
4052 Basel
Switzerland
Tel. +41 61 683 77 34
Fax +41 61 302 89 18
www.mdpi.com

Micromachines Editorial Office
E-mail: micromachines@mdpi.com
www.mdpi.com/journal/micromachines